T0332631

A PRIMER OF INFINITESIMAL ANALYSIS
SECOND EDITION

One of the most remarkable recent occurrences in mathematics is the refounding, on a rigorous basis, of the idea of infinitesimal quantity, a notion that played an important role in the early development of calculus and mathematical analysis. In this new edition, basic calculus, together with some of its applications to simple physical problems, is represented through the use of a straightforward, rigorous, axiomatically formulated concept of 'zero-square', or 'nilpotent' infinitesimal – that is, a quantity so small that its square and all higher powers can be set, literally, to zero. The systematic employment of these infinitesimals reduces the differential calculus to simple algebra and, at the same time, restores to use the 'infinitesimal' methods figuring in traditional applications of the calculus to physical problems – a number of which are discussed in this book. This edition also contains some additional applications to physics.

John L. Bell is Professor of Philosophy at the University of Western Ontario. He is the author of seven other books, including *Models and Ultraproducts* with A. B. Slomson, *A Course in Mathematical Logic* with M. Machover, *Logical Options* with D. DeVidi and G. Solomon, *Set Theory: Boolean-Valued Models and Independence Proofs*, and *The Continuous and the Infinitesimal in Mathematics and Philosophy.*

'This might turn out to be a boring, shallow book review: I merely LOVED the book. . . the explanations are so clear, so considerate; the author must have taught the subject many times, since he anticipates virtually every potential question, concern, and misconception in a student's or reader's mind.'

– Marion Cohen, *MAA Reviews*

'The book will be of interest to philosophically orientated mathematicians and logicians.'

– *European Mathematical Society*

'John Bell has done a first rate job in presenting an elementary introduction to this fascinating subject. . . . I recommend it highly.'

– J. P. Mayberry, *British Journal for the Philosophy of Science*

A PRIMER OF INFINITESIMAL ANALYSIS SECOND EDITION

JOHN L. BELL
University of Western Ontario

CAMBRIDGE
UNIVERSITY PRESS

University Printing House, Cambridge CB2 8BS, United Kingdom

One Liberty Plaza, 20th Floor, New York, NY 10006, USA

477 Williamstown Road, Port Melbourne, VIC 3207, Australia

314-321, 3rd Floor, Plot 3, Splendor Forum, Jasola District Centre, New Delhi - 110025, India

79 Anson Road, #06-04/06, Singapore 079906

Cambridge University Press is part of the University of Cambridge.

It furthers the University's mission by disseminating knowledge in the pursuit of education, learning and research at the highest international levels of excellence.

www.cambridge.org
Information on this title: www.cambridge.org/9780521887182

First published 2008

A catalogue record for this publication is available from the British Library

Library of Congress Cataloging in Publication data
Bell, J. L. (John Lane)
A Primer of infinitesimal analysis / John L. Bell. – 2nd ed.
p. cm.
Includes bibliographical references and index.
ISBN-13: 978-0-521-88718-2 (hardback)
ISBN-10: 978-0-521-88718-6 (hardback)
1. Nonstandard mathematical analysis. I. Title.
QA299.82.B45 2008
515 – dc22 2007035724

ISBN 978-0-521-88718-2 Hardback

Once again, to Mimi

Contents

Preface

A remarkable recent development in mathematics is the refounding, on a rigorous basis, of the idea of *infinitesimal quantity*, a notion which, before being supplanted in the nineteenth century by the limit concept, played a seminal role within the calculus and mathematical analysis. One of the most useful concepts of infinitesimal to have thus acquired rigorous status is that of a quantity so small (but not actually zero) that its square and all higher powers can be set to zero. The introduction of these 'zero-square' or 'nilpotent' infinitesimals opens the way to a revival of the intuitive, and remarkably efficient, 'pre-limit' approaches to the calculus: this little book is an attempt to get the process going at an elementary level. It begins with a historico-philosophical introduction in which the leading ideas of the basic framework – that of *smooth infinitesimal analysis* or analysis in *smooth worlds* – are outlined. The first chapter contains an axiomatic description of the essential technical features of smooth infinitesimal analysis. In the chapters that follow, nilpotent infinitesimals are used to develop single- and multi-variable calculus (with applications), the definite integral, and Taylor's theorem. The penultimate chapter contains a brief introduction to *synthetic differential geometry* – the transparent formulation of the differential geometry of manifolds made possible in smooth infinitesimal analysis by the presence of nilpotent infinitesimals. In the final chapter we outline the novel logical features of the framework. Scattered throughout the text are a number of straightforward exercises which the reader is encouraged to solve.

My purpose in writing this book has been to show how the traditional infinitesimal methods of mathematical analysis can be brought up to date – restored, so to speak – allowing their beauty and utility to be revealed anew. I believe that the greater part of its contents will be intelligible – and rewarding – to anyone with a basic knowledge of the calculus.*

* The only exception to this occurs in Chapters 7 and 8, and the Appendix (all of which can be omitted at a first reading) whose readers are assumed to have a slender acquaintance with differential geometry, logic, and category theory, respectively.

A final remark: The theory of infinitesimals presented here should not be confused with that known as *nonstandard analysis*, invented by Abraham Robinson in the 1960s. The infinitesimals figuring in his formulation are '*invertible*' (arising, in fact, as the 'reciprocals' of *infinitely large* quantities), while those with which we shall be concerned, being nilpotent, cannot possess inverses. The two theories also have quite different mathematical origins, nonstandard analysis arising from developments in logic, and that presented here from category theory. For a brief discussion of nonstandard analysis, see the final chapter of the book.

In this second edition of the book, I have added some new material and taken the opportunity to correct a number of errors.

Acknowledgements

My thanks go to F. W. Lawvere for his helpful comments on an early draft of the book and for his staunch support of the idea of a work of this kind. I would also like to record my gratitude to Roger Astley of Cambridge University Press for his unfailing courtesy and efficiency.

I am also grateful to Thomas Streicher for his careful reading of the first edition and for pointing out a number of errors.

Introduction

According to the *Encyclopedia Britannica* (11th edition, 1913, Volume 14, p. 535, emphasis added),

> The *infinitesimal calculus* is the body of rules and processes by which *continuously varying magnitudes* are dealt with in mathematical analysis. The name '*infinitesimal*' has been applied to the calculus because most of the leading results were first obtained by means of arguments about 'infinitely small' quantities; the 'infinitely small' or 'infinitesimal' quantities were vaguely conceived as being neither zero nor finite but in some intermediate, nascent or evanescent state.

In this passage attention has been drawn to two important, and closely related, mathematical concepts: *continuously varying magnitude* and *infinitesimal*. The first of these is founded on the traditional idea of a *continuum*, that is to say, the domain over which a continuously varying magnitude *actually varies*. The characteristic features of a (connected) continuum are, first, that *it has no gaps* – it 'coheres' – so that a magnitude varying over it has no 'jumps' and, secondly, that it is *indefinitely divisible*. Thus it has been held by a number of prominent thinkers that continua are *nonpunctate*, that is, not 'composed of' or 'synthesized from' discrete points. Witness, for example, the following quotations:

> *Aristotle*: ... no continuum can be made up out of indivisibles, granting that the line is continuous and the point indivisible.
>
> (*Aristotle, 1980, Book 6, Chapter 1*)

> *Leibniz*: A point may not be considered a part of a line.
>
> (*Quoted in Rescher, 1967, p. 109*)

> *Kant*: Space and time are *quanta continua* ... points and instants mere positions ... and out of mere positions viewed as constituents capable of being given prior to space and time neither space nor time can be constructed.
>
> (*Kant, 1964, p. 204*)

Poincaré: . . . between the elements of the continuum [there is supposed to be] a sort of intimate bond which makes a whole of them, in which the point is not prior to the line, but the line to the point.

(*Quoted in Russell, 1937, p. 347*)

Weyl: Exact time- or space-points are not the ultimate, underlying, atomic elements of the duration or extension given to us in experience.

(*Weyl, 1987, p. 94*)

A true continuum is simply something connected in itself and cannot be split into separate pieces; that contradicts its nature.

(*Weyl, 1921: quoted in van Dalen, 1995, p. 160*)

Brouwer: The linear continuum is not exhaustible by the interposition of new units and can therefore never be thought of as a mere collection of units.

(*Brouwer, 1964, p. 80*)

René Thom: . . . a true continuum has no points.

(*See Cascuberta and Castellet (eds), 1992, p. 102*)

We note that these views are much at variance with the generally accepted set-theoretical formulation of mathematics in which all mathematical entities, being synthesized from collections of individuals, are ultimately of a *discrete* or *punctate* nature. This punctate character is possessed in particular by the set supporting the 'continuum' of real numbers – the 'arithmetical continuum'. As applied to the arithmetical continuum 'continuity' is accordingly not a property of the collection of real numbers *per se*, but derives rather from certain features of the additional structures – order-theoretic, topological, analytic – that are customarily imposed on it.

Closely associated with the concept of continuum is the second concept, that of 'infinitesimal'. Traditionally, an infinitesimal quantity is one which, while not necessarily coinciding with zero, is in some sense smaller than any finite quantity. In 'practical' approaches to the differential calculus an infinitesimal quantity or number is one so small that its square and all higher powers can be neglected, i.e. set to zero: we shall call such a quantity a *nilsquare infinitesimal*. It is to be noted that the property of being a nilsquare infinitesimal is an *intrinsic* property, that is, in no way dependent on comparisons with other magnitudes or numbers. An infinitesimal magnitude may be regarded as what remains after a (genuine) continuum has been subjected to an exhaustive analysis, in other words, as a continuum 'viewed in the small'. In this sense an infinitesimal[1] may be taken to be an 'ultimate part' of a continuum: in this same sense,

[1] Henceforth, the term 'infinitesimal' will mean 'infinitesimal quantity', 'infinitesimal number' or 'infinitesimal magnitude', and the context allowed to determine the intended meaning.

mathematicians have on occasion taken the 'ultimate parts' of curves to be infinitesimal straight lines.

We observe that the 'coherence' of a genuine continuum entails that any of its (connected) parts is also a continuum, and accordingly, divisible. A point, on the other hand, is by its nature not divisible, and so (as asserted by Leibniz in the quotation above) cannot be part of a continuum. Since an infinitesimal in the sense just described is a part of the continuum from which it has been extracted, it follows that it cannot be a point: to emphasize this we shall call such infinitesimals *nonpunctiform*.

Infinitesimals have a long and somewhat turbulent history. They make an early appearance in the mathematics of the Greek atomist philosopher Democritus (*c.* 450 BC), only to be banished by the mathematician Eudoxus (*c.* 350 BC) in what was to become official 'Euclidean' mathematics. Taking the somewhat obscure form of 'indivisibles', they reappear in the mathematics of the late middle ages and were systematically exploited in the sixteenth and seventeenth centuries by Kepler, Galileo's student Cavalieri, the Bernoulli clan, and others in determining areas and volumes of curvilinear figures. As 'linelets' and 'timelets' they played an essential role in Isaac Barrow's 'method for finding tangents by calculation', which appears in his *Lectiones Geometricae* of 1670. As 'evanescent quantities' they were instrumental in Newton's development of the calculus, and as 'inassignable quantities' in Leibniz's. De l'Hospital, the author of the first treatise on the differential calculus (entitled *Analyse des Infiniment Petits pour l'Intelligence des Lignes Courbes*, 1696) invokes the concept in postulating that 'a curved line may be regarded as made up of infinitely small straight line segments' and that 'one can take as equal two quantities which differ by an infinitely small quantity'. Memorably derided by Berkeley as 'ghosts of departed quantities' and roundly condemned by Bertrand Russell as 'unnecessary, erroneous, and self-contradictory', these useful, but logically dubious entities were believed to have been finally supplanted by the limit concept which took rigorous and final form in the latter half of the nineteenth century. By the beginning of the twentieth century, most mathematicians took the view that – in analysis at least – the concept of infinitesimal had been thoroughly exploded.

Now in fact, the proscription of infinitesimals did not succeed in eliminating them altogether but, instead, drove them underground. Physicists and engineers, for example, never abandoned their use as a heuristic device for deriving (correct!) results in the application of the calculus to physical problems. And differential geometers as reputable as Lie and Cartan relied on their use in formulating concepts which would later be put on a 'rigorous' footing. And, in a technical sense, they lived on in algebraists' investigations of non-archimedean fields. The concept of infinitesimal even managed to retain some public champions, one of the most active of whom was the philosopher–mathematician Charles Sanders Peirce, who saw the concept of the continuum (as did Brouwer) as arising from the subjective grasp of the flow of time and the subjective 'now' as a nonpunctiform infinitesimal. Here are a few of his observations on these matters:

> It is singular that nobody objects to $\sqrt{-1}$ as involving any contradiction, nor, since Cantor, are infinitely great quantities objected to, but still the antique prejudice against infinitely small quantities remains.
>
> (*Peirce, 1976, p. 123*)

> It is difficult to explain the fact of memory and our apparently perceiving the flow of time, unless we suppose immediate consciousness to extend beyond a single instant. Yet if we make such a supposition we fall into grave difficulties, unless we suppose the time of which we are immediately conscious to be strictly infinitesimal.
>
> (*ibid., p. 124*)

> [The] continuum does not consist of indivisibles, or points, or instants, and does not contain any except insofar as its continuity is ruptured.
>
> (*ibid., p. 925*)

In recent years, the concept of infinitesimal has been refounded on a solid basis. First, in the 1960s Abraham Robinson, using methods of mathematical logic, created *nonstandard analysis*, in which Leibniz's infinitesimals – conceived essentially as infinitely small but nonzero real numbers – were finally incorporated into the real number system without violating any of the usual rules of arithemetic (see Robinson, 1966). And in the 1970s startling new developments in the mathematical discipline of category theory led to the creation of *smooth infinitesimal analysis*, a rigorous axiomatic theory of nilsquare and nonpunctiform infinitesimals. As we show in this book, within smooth infinitesimal analysis the basic calculus and differential geometry can be developed along traditional 'infinitesimal' lines – with full rigour – using straightforward calculations with infinitesimals in place of the limit concept.

Just as with nonEuclidean geometry, the consistency of smooth infinitesimal analysis is established by the construction of various *models* for it[2]. Each model is a mathematical structure (a category) of a certain kind containing all the usual geometric objects such as the real line and Euclidean spaces, together with transformations or maps between them. Their key feature is that within each all maps between geometric objects are *smooth*[3] and *a fortiori*

[2] For a sketch of the construction of these models, see the Appendix.

[3] A map between two mathematical objects each supporting a differential structure is said to be smooth if it is differentiable arbitrarily many times. In particular, a smooth map and all its derivatives must be continuous.

continuous[4]. For this reason, any one of these models of smooth infinitesimal analysis will be referred to as a *smooth world*; we shall sometimes use the symbol $\mathbb{S}$ to denote an arbitrary smooth world.

Now in order to achieve universal continuity of maps within smooth worlds, and thereby to ensure the consistency of smooth infinitesimal analysis, it turns out that a certain logical price must be paid. In fact, one is forced to acknowledge that the so-called *law of excluded middle* – every statement is either definitely true or definitely false – cannot be generally affirmed within smooth worlds[5]. This stems from the fact that unconstrained use of the law of excluded middle legitimizes the construction of *discontinuous* functions, as the following simple argument shows. Assuming the law of excluded middle, each real number is either equal to 0 or unequal to 0, so that correlating 1 to 0 and 0 to each nonzero real number defines a function – the 'blip' function – on the real line which is obviously discontinuous. So, if the law of excluded middle held in a smooth world $\mathbb{S}$, the discontinuous blip function could be defined there (see Fig. 1). Thus, since all functions in $\mathbb{S}$ are continuous, it follows that the law of

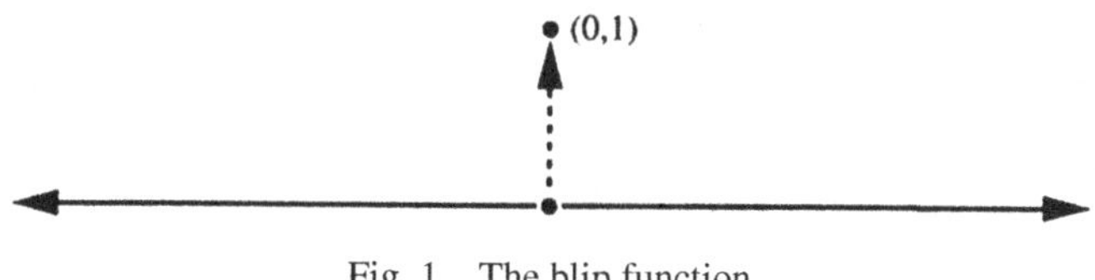

Fig. 1 The blip function

excluded middle must fail within it. More precisely, this argument shows that the statement

$$\text{for any real number } x, \text{ either } x = 0 \text{ or } \textit{not}\, x = 0$$

is *false* in $\mathbb{S}$.

Another way of showing that arbitrary statements interpreted in a smooth world cannot be regarded as possessing one of just two 'truth values' *true* or *false* runs as follows. Let Ω be the set of truth values in $\mathbb{S}$ (which we assume contains at least *true* and *false* as members). Then in $\mathbb{S}$, as in ordinary set theory,

[4] Thus each such model may be thought of as embodying Leibniz's doctrine *natura non facit saltus* – nature makes no jump.

[5] As the following quotation shows, Peirce was aware, even before Brouwer, that a faithful account of the truly continuous will involve jettisoning the law of excluded middle:

> Now if we are to accept the common idea of continuity . . . we must either say that a continuous line contains no points or . . . that the principle of excluded middle does not hold of these points. The principle of excluded middle applies only to an individual . . . but places being mere possibilities without actual existence are not individuals.
>
> (*Peirce, 1976, p. xvi: the quotation is from a note written in 1903*)

functions from any given object X to Ω correspond exactly to parts of X, proper nonempty parts corresponding to nonconstant functions. If X is a connected continuum (e.g. the real line), it presumably does have proper nonempty parts but certainly no nonconstant continuous functions to the two element set {*true, false*}. It follows that, in $\mathbb{S}$, the set of truth values cannot reduce to {*true, false*}. Thus logic in smooth worlds is *many-valued* or *polyvalent*.

Essentially the same argument shows that, in a smooth world, a connected continuum X is continuous in the strong sense that its only detachable parts are X itself and its empty part: here a part U of X is said to be *detachable* if there is a *complementary* part V of X such that U and V are disjoint and together cover X. For, clearly, detachable parts of X correspond to maps on X to {*true, false*}, so since all such maps on X are constant, and they in turn correspond to X itself and its empty part, these latter are the sole detachable parts of X[6].

Now at first sight in the failure of the law of excluded middle in smooth worlds may seem to constitute a major drawback. However, it is precisely this failure which allows nonpunctiform infinitesimals to be present. To get some idea of why this is so, we observe that since the law of excluded middle fails in any smooth world $\mathbb{S}$, so does its logical equivalent the *law of double negation*: for any statement A, *not not* A implies A. If we now call two points a, b on the real line *distinguishable* or *distinct* when they are not identical, i.e. *not* $a = b$ – which as usual we shall write $a \neq b$ – and indistinguishable in the contrary case, i.e. if *not* $a \neq b$, then, in $\mathbb{S}$, indistinguishability of points will not in general imply their identity. As a result, the 'infinitesimal neighbourhood of 0' comprising all points indistinguishable from 0 – which we will denote by I – will, in $\mathbb{S}$, be nonpunctiform in the sense that it does not reduce to {0}, that is,

it is not the case that 0 is the sole member of I.

If we call the members of I *infinitesimals*, then this statement may be rephrased:

it is not the case that all infinitesimals coincide with 0.

Observe, however, that we evidently cannot go on to infer from this that

there exists an infinitesimal which is $\neq 0$.

[6] In this connection it is worth drawing attention to the remarkable observation of Weyl (1940), who realized that the essential nature of continua can only be given full expression within a context resembling our smooth worlds:

> A natural way to take into account the nature of a continuum which, following Anaxagoras, defies 'chopping off its parts with a hatchet' would be by limiting oneself to continuous functions.

For such an entity would possess the property of being both distinguishable and indistinguishable from 0, which is clearly impossible[7]. What this means is that, while in $\mathbb{S}$, it would be incorrect to assume that all infinitesimals coincide with 0, it would be no less incorrect to suppose that we can single out an actual nonzero infinitesimal, i.e. one which is distinguishable from 0. In other words, nonzero infinitesimals can, and will, be present only in a 'virtual' sense[8]. Nevertheless, as we shall see, this virtual existence will suffice for the development of 'infinitesimal' analysis in smooth worlds.

In traditional mathematics two distinct, but closely related, conceptions of nonpunctiform infinitesimal can be discerned. Both may be considered as resulting from the attempt to measure continua in terms of discrete entities. The first of these conceptions stems from the idea that, just as the perimeter of a polygon is the sum of its finite discrete collection of edges, so any continuous curve should be representable as the 'sum' of an (infinite) discrete collection of infinitesimally short linear segments – the 'linear infinitesimals' of the curve. This conception was formulated by l'Hospital, and also advanced in some form two millenia earlier by the Greek mathematicians Antiphon and Bryson (*c.* 450 BC)[9]. The second concept arises analogously from the idea that a continuous surface or volume can be conceived as the sum of an indefinitely large, but discrete assemblage of lines or planes, the so-called *indivisibles*[10] of the surface or volume. This idea, exploited by Cavalieri in the seventeenth century, also appears in Archimedes' *Method*.

Let us show, by means of an example, how these two concepts of infinitesimal are related, and how they give rise to the concept of nilsquare infinitesimal. Given a smooth curve AB, suppose we want to evaluate the area of the region $ABCO$ by regarding it as the sum of thin rectangles $XYRS$ (Fig. 2). If X and S are distinguishable points then so are Y and R, so that the 'area defect' ∇ under the curve is nonzero; in this event the figure $ABCO$ cannot literally be the sum of such rectangles as $XYRS$. On the other hand, if X and S coincide, then ∇ is zero but $XYRS$ collapses into a straight line, thus failing altogether

[7] Although the law of excluded middle has had to be abandoned, the law of noncontradiction – a statement and its negation cannot both be true – will of course continue to be upheld in $\mathbb{S}$.

[8] The virtual infinitesimals of smooth worlds resemble both the virtual displacements of classical dynamics and the virtual particles of contemporary particle physics. Each has no more than only a transitory presence, and vanishes at the completion of a calculation (in the first two cases) or an interaction (in the last case).

[9] See Boyer (1959), Chapter II.

[10] The use of this term in connection with continua, although traditional, is a trifle unfortunate since no part of a continuum is 'indivisible'. This fact seems to have contributed to the general confusion – which I hope is not compounded here – surrounding the notion. See Boyer (1959), especially Chapter III.

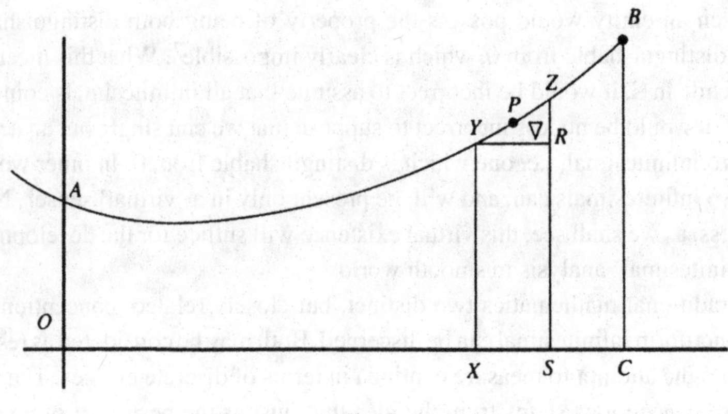

Fig. 2

to contribute to the area of the figure. In order, therefore, for *ABCO* to be the sum of rectangles like *XYRS*, we require that their base vertices *X, S* be indistinguishable without coinciding, and yet the area defect ∇ be zero. This desideratum (which is patently incompatible with the law of excluded middle) necessitates that the segment *XS* be a nondegenerate[11] linear infinitesimal of a special kind: let us appropriate Barrow's delightful term and call it a *linelet.*

Now to achieve our object we want *YRZ* to be a nondegenerate triangle of zero area. For this to be the case we clearly require first that

(a) the segment *YZ* of the curve around the point *P* is actually straight and nondegenerate (in particular, does not reduce to *P*).

In this event, the area ∇ of *YRZ* is proportional to the square of the length of the line *XS*, so that, if this area is to be zero, we must further require that

(b) *XS* is nondegenerate of length ε with $\varepsilon^2 = 0$, that is, ε is a nilsquare infinitesimal.

If, for any point *P*, a segment *YZ* of the curve exists such that the corresponding conditions (a) and (b) are satisfied, then the rectangles *XYRS* may be regarded as indivisibles whose sum exhausts the figure. Accordingly, an indivisible of the figure may be identified as a rectangle with a linelet as base.

[11] Here and throughout we term 'nondegenerate' any figure not identical with a single point.

If this procedure is to be performable for any curve, (a) needs to be extended to the following principle:

I. For any smooth curve C and any point P on it, there is a (small) nondegenerate segment of C – a *microsegment* – around P which is straight, that is, C is *microstraight* around P.

And (b) must be extended to the following principle:

II. The set Δ of magnitudes ε for which $\varepsilon^2 = 0$ – the *nilsquare infinitesimals* – does not reduce to $\{0\}$.

Principle II, which will be instrumental in reducing the differential calculus to simple algebra in our account of smooth infinitesimal analysis, is actually a consequence of Principle I. For, assuming I, consider the curve C with equation $y = x^2$ (Fig. 3). Let U be the straight portion of the curve around the origin: U

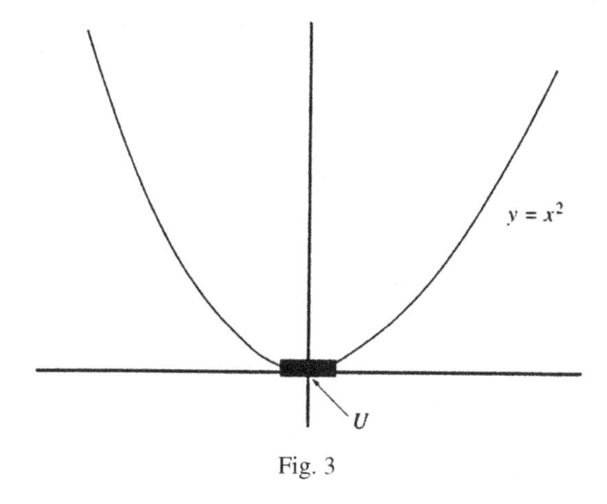

Fig. 3

is the intersection of the curve with its tangent at the origin (the x-axis). Thus U is the set of points x on the real line satisfying $x^2 = 0$. In other words, U and Δ are identical. Since I asserts the nondegeneracy of U, that is, of Δ, we obtain II.

Principle I, which we shall term the *Principle of Microstraightness* (for smooth curves) – and which will play a key role in smooth infinitesimal analysis – is closely related both to *Leibniz's Principle of Continuity*, and to what we shall call the *Principle of Microuniformity* (of natural processes). Leibniz's principle, in essence, is the assertion that processes in nature occur continuously, while the Principle of Microuniformity is the assertion that any such process may be considered as taking place at a constant rate over any sufficiently small period of time (i.e. over Barrow's 'timelets'). For example, if the process is the motion of a particle, the Principle of Microuniformity entails that over a timelet the particle experiences no accelerations. This idea, although rarely explicitly stated, is freely employed in a heuristic capacity in classical mechanics and the theory of differential equations. We observe in passing that the Principle of Continuity is actually a consequence of the Principle of Microuniformity.

The close relationship between the Principles of Microuniformity and Microstraightness becomes manifest when natural processes – for example, the motions of bodies – are represented as curves correlating dependent and independent variables. For then, microuniformity of the process is represented by microstraightness of the associated curve.

The Principle of Microstraightness yields an intuitively satisfying account of *motion*. For it entails that infinitesimal parts of (the curve representing) a motion are not degenerate 'points' where, as Aristotle observed millenia ago, no motion is detectable (or, indeed, even possible!), but are, rather, nondegenerate spatial segments just large enough to make motion over each one palpable. On this reckoning, states of motion are to be taken seriously, and not merely identified with their result: the successive occupation of a series of distinct positions. Instead, a state of motion is represented by the smoothly varying straight microsegment of its associated curve. This straight microsegment may be thought of as an infinitesimal 'rigid rod', just long enough to have a slope – and so, like a speedometer needle, to indicate the presence of motion – but too short to bend. It is thus an entity possessing (location and) *direction without magnitude*, intermediate in nature between a point and a Euclidean straight line.

This analysis may also be applied to the mathematical representation of *time*. Classically, time is represented as a succession of discrete instants, isolated 'nows', where time has, as it were, stopped. The Principle of Microstraightness, however, suggests rather that time be regarded as a plurality of smoothly overlapping timelets each of which may be held to represent a 'now' (or 'specious present') and over which time is, so to speak, still passing. This conception of the nature of time is similar to that proposed by Aristotle (*Physics*, Book 6, Chapter ix) to refute Zeno's paradox of the arrow.

Most important for our purposes, however, the Principle of Microstraightness decisively solves the problem of assigning a quantitative meaning to the concept of *instantaneous rate of change* – the fundamental concept of the differential calculus. For, given a smooth curve representing a physical process, the instantaneous rate of change of the process at a point P on the curve is given simply by the slope of the straight microsegment ℓ forming part of the curve at P: ℓ is of course part of the tangent to the curve at P. If the curve has equation $y = f(x)$ and P has coordinates (x_0, y_0), then the slope of the tangent to the curve is given, as usual, by the value $f'(x_0)$ of the *derivative* f' of f at x_0. The presence of nilsquare infinitesimals guaranteed by the Principle of Microstraightness enables this value to be calculated in a straightforward manner, as is shown by the following informal argument (Fig. 4).

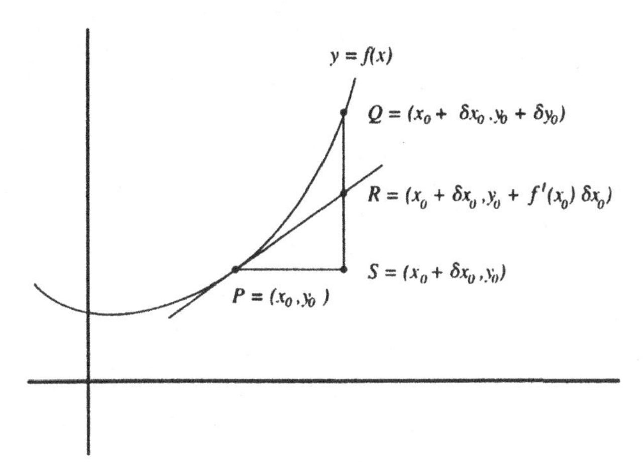

Fig. 4

Let δx_0 be any small change in x_0. The corresponding small change δy_0 in $y_0 = f(x_0)$, represented by the line *SQ* in Fig. 4, may then be split into two components. The first of these is the change in y_0 along the tangent to the curve at *P*, and is represented by the line *SR*, which has length $f'(x_0)\delta x_0$. The second is represented by the line *QR*; we write its length in the form $H(\delta x_0)^2$, where H is some quantity depending both on x_0 and δx_0. Thus

$$f(x_0 + \delta x_0) - f(x_0) = \delta y_0 = f'(x_0)\delta x_0 + H(\delta x_0)^2.$$

Now suppose that δx_0 is a nilsquare infinitesimal ε. Then $(\delta x_0)^2 = \varepsilon^2 = 0$[12] and so the equation above reduces to

$$f(x_0 + \varepsilon) - f(x_0) = f'(x_0)\varepsilon.$$

Allowing x_0 to vary, we see then that the value of $f(x)$ attendant upon a nilsquare infinitesimal change ε in x is exactly equal to $f'(x)\varepsilon$. The derivative $f'(x)$ is thus determined as that quantity A satisfying the equation

$$f(x + \varepsilon) - f(x) = A\varepsilon$$

for all nilsquare infinitesimal ε.

[12] Notice that then the length $H(\delta x_0)^2$ of *QR* is zero. Thus *Q* and *R* coincide and so, in accordance with the Principle of Microstraightness, the portion *PQ* of the curve coincides with the portion *PR* of the tangent and is therefore 'straight'.

The axioms of smooth infinitesimal analysis will permit us to define the derivative $f'(x)$ to be the unique quantity A satisfying this last equation for all nilsquare infinitesimal ε. As we shall see, defining the derivative in this way enables the basic rules and processes of the differential calculus to be reduced to simple algebra.

In this book our main purpose will be to develop mathematics within smooth infinitesimal analysis, and so we shall not be directly concerned with the technical construction of its models as categories. Here we give just a bare outline of the construction: the Appendix contains a further sketch and full details may be found in Moerdijk and Reyes (1991).

Roughly speaking, a *category* is a mathematical system whose basic constituents are not only mathematical 'objects' (in set theory these are the 'sets'), but also 'maps' ('functions', 'transformations', 'correlations') between the said objects. By contrast with set theory, in a category the 'maps' have an autonomous character which renders them in general not definable in terms of the objects, so that one has a great deal of freedom in deciding exactly what these maps should be. A crucial feature of maps in a category is that each map f is associated with a specific pair of objects written $\mathrm{dom}(f)$, $\mathrm{cod}(f)$ and called its *domain* and *codomain*, respectively. We think of any map as being defined on its domain and taking values in its codomain, or as going from its domain to its codomain: to indicate this we employ the customary notation $f: A \to B$, where f is any map and $A = \mathrm{dom}(f)$, $B = \mathrm{cod}(f)$. Another basic feature of maps in a category is that certain pairs of them can be *composed* to yield new maps. To be precise, associated with each pair of maps $f: A \to B$ and $g: B \to C$ such that $\mathrm{cod}(f) = \mathrm{dom}(g)$ (f is then said to be composable with g) is a map $g \circ f: A \to C$ called its composite: it is supposed that composition is associative in the sense that, if (f, g) and (g, h) are composable pairs of maps, then $h \circ (g \circ f) = (h \circ g) \circ f$. Finally, associated with each object A is a map $1_A: A \to A$ called the *identity map* on A; it is assumed that, for any $f: B \to A$ and $g: A \to C$ we have $1_A \circ f = f$ and $g \circ 1_A = g$. Possession of these three properties actually defines the notion of category. Two prominent examples of categories are **Set**, the *category of sets*, with all ordinary sets as objects and all functions between them as maps, and **Man**, the *category of* (smooth) *manifolds*, with all smooth manifolds as objects and all smooth functions between them as maps.

Now there is a certain sort of category possessing an internal structure sufficiently rich to enable all of the usual constructions of mathematics to be carried out. Categories of this sort are called *toposes*; **Set** (but not **Man**) is a topos[13].

[13] Without giving a formal definition of a topos, we may say that it is a category **E** which resembles **Set** in the following respects: (1) it contains an object 1 which behaves like a one-element set;

Toposes may be suggestively described as 'universes of discourse' within which the objects are undergoing *variation* or *change* in some way: the category of sets is a topos in which the variation of the objects has been reduced to zero, the static, timeless case[14]. Associated with each such 'universe of discourse' is a mathematical language – a formal version of the familiar language used in set theory – which serves to 'chart' that universe, and which contains, in coded form, a complete description of it. Just as all the charts in an atlas share a common geometry, so all the formal languages associated with toposes or 'universes of discourse' share a common logic. This logic turns out to be what is known as *constructive* or *intuitionistic logic*[15], in which existential propositions can be affirmed only when the term whose existence is asserted can be constructed or named in some definite way, and in which a disjunction can be affirmed only when a definite one of the disjuncts has been affirmed. Roughly speaking, in constructive logic, all the principles of classical logic are affirmed with the exception of those that depend for their validity on the law of excluded middle[16]. (Recall that this law fails in smooth worlds.) If the general principles of constructive reasoning are adhered to – and all this means in practice is avoiding certain 'arguments by contradiction'[17] – then mathematical arguments within these 'universes of discourse' can take essentially the same form as they do in 'ordinary' mathematics.

From the category-theoretic point of view, furnishing mathematical analysis and differential geometry with a 'set-theoretic' foundation amounts to embedding **Man** in **Set**: the latter's stronger properties – it is a topos, **Man** is not – then sanction the performance of necessary constructions (notably, the formation of

(2) any pair of objects A, B determines an object $A \times B$ which behaves like the Cartesian product of A and B; (3) any pair of objects A, B determines an object A^B which behaves like the 'set of all maps' from B to A; (4) it contains an object Ω playing a role similar to that played in **Set** by the 'truth value' set $2 = \{0, 1\}$: maps from an arbitrary object A to Ω correspond bijectively to 'subobjects' of A. For some further details see the Appendix.

14 The simplest example of a topos in which genuine 'variation' is taking place is the category $\mathbf{Set}^2$ of sets varying over two moments of time 0 ('then') and 1 ('now'). The objects of $\mathbf{Set}^2$ are triples (f, A_0, A_1) where $f, A_0 \to A_1$ is a map in **Set**: we think of each such triple as a 'varying set' in which A_0 was its state 'then', A_1 its state 'now' and f is the transition function between the two states. A map in $\mathbf{Set}^2$ between two such triples (f, A_0, A_1) and (g, B_0, B_1) is a pair h_0: $A_0 \to B_0$, h_1: $A_1 \to B_1$ in **Set** which preserve the transition functions f and g, i.e. for which the composites $g \circ h_0$ and $h_1 \circ f$ coincide. More generally, the simple structure $2 = \{0, 1\}$ may be replaced by an arbitrary category **C** to yield the category $\mathbf{Set}^{\mathbf{C}}$ of sets varying over **C**. For details see, for example, Bell (1988b).

15 See Chapter 8 for a description of the rules of constructive logic.

16 By this we do not mean that the law of excluded middle is explicitly denied (i.e. that its negation is derivable) in constructive logic, only that, as we have said, it is not affirmed. Because of this, classical logic may be regarded as the special or idealized version of constructive logic in which the law of excluded middle is postulated. And of course there are toposes, notably **Set**, whose associated logic is classical, i.e. the law of excluded middle holds there.

17 To be precise, those *reductio ad absurdum* arguments that derive a proposition from the absurdity of its denial.

tangent spaces) which cannot be carried out directly in **Man**. However, in the process of embedding **Man** in **Set** we obtain, not only new objects (i.e. pure sets which are not correlated with manifolds), but also new (discontinuous) maps between the old objects. (For example, the blip function considered earlier appears in **Set** but not in **Man**.) Moreover, despite the (inevitable) presence of many new objects in **Set**, none of them can play the role of 'infinitesimal' objects such as Δ or I above[18]. By contrast, in constructing a smooth world we seek to embed **Man** in a topos **E** which does not contain new maps between manifolds (so that all such maps in **E** are still smooth), yet does contain 'infinitesimal' objects: in particular, an object Δ which realizes the Principle of Microstraightness. Maps in **E** with domain Δ may then be identified with 'straight microsegments' of curves in the sense introduced above.

Recent work has shown that toposes – the so-called *smooth toposes* – can be constructed so as to meet these requirements (and also to satisfy the additional principles to be introduced in the sequel). These toposes are obtained by embedding the category of manifolds in an enlarged category **C** which contains 'infinitesimal' objects, and forming the topos $\mathbf{Set}^{\mathbf{c}}$ of sets 'varying over' **C**. Each smooth topos **E** is then identified as a certain subcategory of $\mathbf{Set}^{\mathbf{c}}$. Any one of these toposes has the property that its objects are undergoing a form of smooth variation, and each may be taken as a smooth world. Smooth infinitesimal analysis – mathematics in smooth worlds – can then, as in any topos, be developed in the straightforward informal style of 'ordinary' mathematics (a procedure to be adopted in this book).

These facts guarantee the consistency of smooth infinitesimal analysis, and so also the essential soundness of (many of) the infinitesimal methods employed by the mathematicians of the past. This is a striking achievement, since the conception of infinitesimal supporting these methods was vague and occasionally gave rise to outright inconsistencies. Now it may be plausibly maintained that such inconsistencies ultimately arose from the fact that infinitesimals, as intrinsically varying[19] quantities, are logically incompatible – at least, within the canons of classical logic – with the static quantities traditionally employed in mathematics. So it would seem natural to attempt to eradicate this incompatibility by allowing the static quantities themselves to vary continuously in a manner consonant with the variation of their infinitesimal counterparts. In the smooth worlds constructed within category theory this goal is achieved, in essence, by

[18] This is because the presence of objects like I or Δ leads to the failure of the law of excluded middle, which, as we have observed, holds in **Set**.

[19] This is a consequence of their being in a 'nascent or evanescent state'.

ensuring that all quantities – infinitesimal and 'static' alike – are undergoing smooth variation. At the same time, the problem of vagueness of the concept of infinitesimal is overcome through the device of furnishing every quantity – and in particular every infinitesimal quantity – with a definite domain over which it varies and a definite codomain in which it takes values. The presence of non-punctiform infinitesimals happily restores to the continuum concept Poincaré's 'intimate bond' between elements absent in arithmetical or set theoretic formulations. And finally, the necessary failure in the models of the law of excluded middle suggests that it was the unqualified acceptance of the correctness of this law, rather than any inherent logical flaw in the concept of infinitesimal itself, which for so long prevented that concept from achieving mathematical respectability.

In the chapters that follow, we will show how elementary calculus and some of its principal applications can be developed within smooth infinitesimal analysis in a simple algebraic manner, using calculations with nilsquare infinitesimals in place of the classifical limit concept. In Chapters 1 and 2 are described the basic features of smooth worlds and the development of elementary calculus in them. Chapters 3 and 4 are devoted to applications of the differential calculus in smooth infinitesimal analysis to a range of traditional geometric and physical problems. In Chapter 5 we introduce and apply the differential calculus of several variables in smooth infinitesimal analysis. Chapter 6 contains a treatment of the elementary theory of the definite integral in smooth infinitesimal analysis, together with a discussion of higher-order infinitesimals and their uses. In the penultimate chapter we give a brief and elementary introduction to differential geometry in smooth infinitesimal analysis: we will see that the presence of infinitesimals enables the basic constructions to be cast in a form that is simpler and much more intuitive than is possible classically. The final chapter, which is intended for logicians, contains an account of smooth infinitesimal analysis as an axiomatic system, and a comparison with nonstandard analysis. In the Appendix we sketch the construction of models of smooth infinitesimal analysis.

It is hoped that those readers less concerned with technical applications than with the acquisition of a basic grasp of the principles underlying smooth infinitesimal analysis will find that this introduction, conjoined with Chapters 1, 2 and 8, form a self-contained presentation meeting their requirements.

1

Basic features of smooth worlds

The fundamental object in any smooth world $\mathbb{S}$ is an indefinitely extensible homogeneous straight line R – the *smooth, affine* or *real line*. We assume that we are given the notion of a *location* or *point* in R, together with the relation $=$ of *identity* or *coincidence* of locations. We use lower case letters $a, b, \ldots$, $x, y, \ldots, \alpha, \beta, \ldots$ for locations. We write $a \neq b$ for *not* $a = b$: this may be read 'a and b are *distinct* or *distinguishable*'. It is important to be aware (cf. the remarks in the Introduction) that we do not assume that the identity relation on R is decidable in the sense that, for any a, b, either $a = b$ or $a \neq b$: thus we allow for the possibility that locations may not be presented with sufficient definiteness to enable a decision as to their identity or distinguishability to be made.

We assume given two distinct points on R which we will denote by 0 and 1 and call the *zero* and the *unit*, respectively. We also suppose that there is defined on R an operation, denoted by $-$, which assigns, to each point a, a point $-a$ called its *reflection* in 0. We assume that $-$ satisfies $-(-a) = a$ and $-0 = 0$.

For each pair a, b of points we assume given an entity $a\hat{\ }b$ which we shall call the *oriented* (a, b)-segment of R. We suppose that, for any points a, b, c, d, $a\hat{\ }b$ and $c\hat{\ }d$ are identical if and only if $a = c$ and $b = d$. The segment $0\hat{\ }a$ will be denoted by a^* and called simply the *segment* of R of *length* a. Segments may be thought of as oriented linear magnitudes: in particular, for each point a, the segment $(-a)^*$ is to be regarded as the segment a 'pointing in the opposite direction'. The (bijective) correspondence $a \rightsquigarrow a^*$ between points and segments/magnitudes enables us to identify each point a with its corresponding magnitude a^*. We shall accordingly employ the terms 'point' and 'magnitude' synonymously, allowing the context to determine which choice is appropriate.

We suppose that, for any pair of points, a, b, we can form a segment a^*: b^* which we shall think of as the segment obtained by *juxtaposing* a^* and b^* (in that order, and preserving their given orientation). We suppose that a^*: b^* is of the form c^* for some unique point c which, as usual, we call the *sum* of (the magnitudes associated with) a and b and denote by $a + b$. We write $a - b$ for $a + (-b)$. We assume that the resulting operation $+$ has the familiar properties:

$$0 + a = a \qquad a - a = 0 \qquad a + b = b + a \qquad (a + b) + c = a + (b + c).$$

In mathematical terminology, we are supposing that the operation $+$ defines an Abelian group structure on (the points of) R, with neutral element 0.

We assume that in $\mathbb{S}$ we can form the Cartesian powers $R \times R$, $R \times R \times R, \ldots, R^n, \ldots$ of R. R^n is, as usual, homogeneous n-dimensional space, each point of which may be identified as an n-tuple $(a_1, \ldots, a_n)$ of points of R. We shall say that two points $\mathbf{a} = (a_1, \ldots a_n)$, $\mathbf{b} = (b_1, \ldots, b_n)$ are *distinct*, and write $\mathbf{a} \neq \mathbf{b}$, if $a_i \neq b_i$ for some explicit $i = 1, \ldots, n$. That is, distinctness of points in n-dimensional space means distinctness of at least one explicit coordinate.

We suppose that the usual Euclidean constructions of products and inverses of magnitudes can be carried out in $R \times R$. Thus (see Fig. 1.1), given two

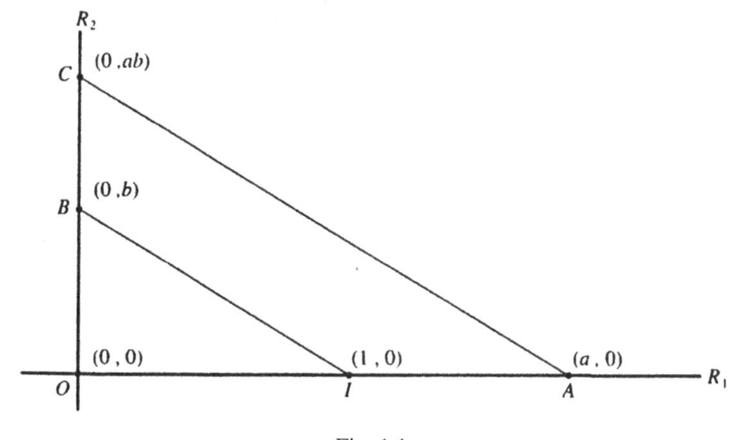

Fig. 1.1

magnitudes a, b, to define their product $a.b$ in $R \times R$ we take two perpendicular copies R_1 (the 'x-axis'), whose points are exactly those of the form $(x, 0)$ and R_2 (the 'y-axis', whose points are exactly those of the form $(0, y)$). R_1 and R_2 intersect at the point $O = (0, 0)$, the 'origin of coordinates'. Now consider the segments OA, OI of lengths a,1, respectively, along R_1 and the segment OB of length b along R_2. The points I and B, being distinct, determine a unique line IB. The line through A parallel to IB intersects R_2 in a point C whose y-coordinate is defined to be the product $a.b$ (which is, more often than not, written ab).

It is important to note that we do not assume that, if $ab = 0$, then either $a = 0$ or $b = 0$. For we do not want to exclude the possibility (which will indeed be realized in $\mathbb{S}$) that a, although not identical with 0, is nonetheless so small that its product with itself is identical with 0.

Given $a \neq 0$, to construct the inverse a^{-1} or $1/a$ we take the x and y axes as before and consider the segments OA, OI along the x-axis of lengths 1, a, respectively, and the segment OB of length 1 along the y-axis (see Fig. 1.2).

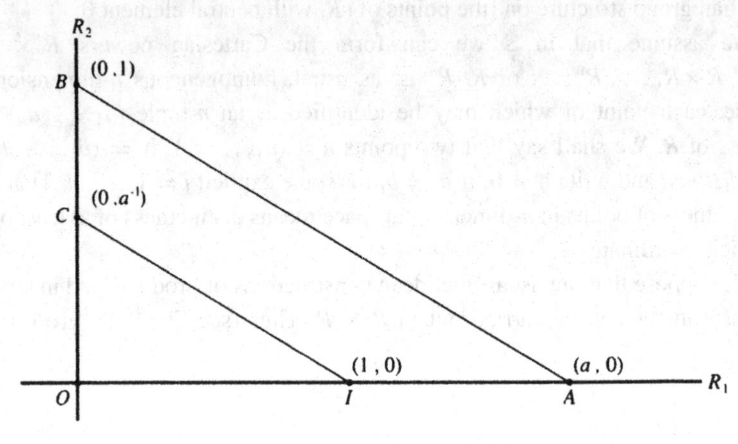

Fig. 1.2

The points A and B, being distinct, determine a unique line AB. The line through I parallel to AB intersects the y-axis in a point C whose y-coordinate is defined to be a^{-1}.

Here it should be noted that, as usual, a^{-1} is defined only when a is distinct from 0.

We assume that products and inverses satisfy the following familiar rules (where we write a/b for $a.b^{-1}$):

$$0.a = 0 \qquad 1.a = a \qquad a.b = b.a \qquad a.(b.c) = (a.b).c$$
$$a.(b + c) = a.c + b.c \qquad a \neq 0 \text{ implies } a/a = 1.$$

In mathematical terminology, (the points of) R, together with the operations of addition (+) and multiplication (·), forms a *field*.

We now suppose that we are given an *order relation* among the points of R which we denote by $<$: $a < b$ (also written $b > a$) is to be understood as asserting that a is strictly to the left of b (or b is strictly to the right of a). We shall assume that $<$ satisfies the following conditions: for any a, b:

(1) $a < b$ and $b < c$ implies $a < c$.
(2) *not* $a < a$.
(3) $a < b$ implies $a + c < b + c$ for any c.
(4) $a < b$ and $0 < c$ implies $ac < bc$.

(5) either $0 < a$ or $a < 1$.
(6) $a \neq b$ implies $a < b$ or $b < a$.

Condition (1) expresses the transitivity of $<$, (2) its strictness, (3) and (4) its compatibility with $+$ and ., and (5) the idea that 1 is sufficiently far to the right of 0 (notice that (5) and (2) jointly imply $0 < 1$) for each point to be either strictly to the right of 0 or strictly to the left of 1. Finally, (6) embodies the idea that, of any two distinguishable points, one is strictly to the left of the other. Notice that (6) does not imply that $<$ satisfies the law of trichotomy, namely, that for any a,b either $a < b$ or $a = b$ or $b < a$. Thus we have automatically allowed for the possibility (which will turn out to be a reality in $\mathbb{S}$!) that two locations, although not in fact coincident, are nonetheless sufficiently indistinguishable that it cannot be decided whether one is to the right or left of the other.

We define the equal to or less than relation $\leq$ on R by

$$a \leq b \text{ if and only if } \textit{not } b < a.$$

The open interval $)a, b($ is defined to consist of those points x for which both $a < x$ and $x < b$, and the closed interval $[a, b]$ to consist of those points x for which both $a \leq x$ and $x \leq b$.

Exercises

1.1 Show that $0 < a$ implies $0 \neq a$; $0 < a$ iff $-a < 0$; $0 < 1 + 1$; and ($a < 0$ or $0 < a$) implies $0 < a^2$.
1.2 Show that, if $a < b$, then, for any x, either $a < x$ or $x < b$.
1.3 Show that $)a, b($ is empty iff *not* $a < b$.
1.4 Show that $\leq$ satisfies the following conditions:

$$x \leq y \text{ and } y \leq z \text{ implies } x \leq z \qquad x \leq x$$
$$x \leq y \text{ implies } x + z \leq y + z$$
$$x \leq y \text{ and } 0 \leq t \text{ implies } xt \leq yt \qquad 0 \leq 1.$$

1.5 Show that any closed interval is *convex* in the sense that, if x and y are in it, so is $x + t(y - x)$ for any t in $[0,1]$.

We also assume that, in $\mathbb{S}$, the extraction of square roots of positive quantities can be performed: that is, we assume the truth in R of the following assertion:

$$\text{for any } a > 0, \text{ there exists } b \text{ such that } b^2 = a.$$

This is tantamount to supposing that the usual Euclidean construction of the square root of a segment can be carried out in $\mathbb{S}$: if a segment of length $a > 0$

is given, mark out a straight line OA of length a and AB of length 1 (see Fig. 1.3). Draw a circle with the segment OB as diameter and construct the perpendicular

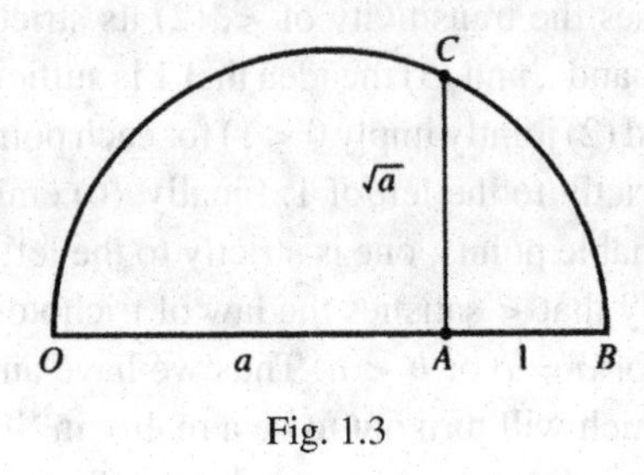

Fig. 1.3

to OB through A, which meets the circle in C (since $a > 0$, A is distinct from O so C is well defined). Then AC has length $\sqrt{a} = a^{\frac{1}{2}}$.

We recall that in our description of $\mathbb{S}$ we have not excluded the possibility that, in R, $a^2 = 0$ can hold without our being able to affirm that $a = 0$. That is, if we define the part Δ of R to consist of those points x for which $x^2 = 0$, or in symbols,

$$\Delta = \{x: x^2 = 0\},$$

it is possible that Δ does not reduce to $\{0\}$: we shall, in fact, shortly adopt a principle which explicitly ensures that this is the case in $\mathbb{S}$.

Henceforth we shall use letters $\varepsilon, \eta, \zeta, \xi$ (possibly with subscripts) as variables ranging over Δ: these will also be referred to as *infinitesimal quantities* or *microquantities*. Δ will also be called the (*basic*) *microneighbourhood* (of 0). We shall say that a part A of R is stable under the addition of microquantities, or *microstable*, if $a + \varepsilon$ is in A whenever a is in A and ε is in Δ.

Exercises

1.6 Show that, for all ε in Δ, (i) *not* ($\varepsilon < 0$ or $0 < \varepsilon$), (ii) $0 \leq \varepsilon$ and $\varepsilon \leq 0$, (iii) for any a in R, εa is in Δ, (iv) if $a > 0$, then $a + \varepsilon > 0$.

1.7 Show that for any a, b in R and all ε, η in Δ, $[a, b] = [a + \varepsilon, b + \eta]$. Deduce that $[a, b]$ is microstable.

We now suppose that the notion of a *function* (also called *map* or *mapping*) between any pair of objects of $\mathbb{S}$ is given. We adopt the usual notation $f: X \to Y$ to indicate that f is a function defined on X with values in Y: X is called the *domain*, and Y the *codomain*, of f. When the domain, codomain and values $f(x)$ of a function f are already known, we shall sometimes introduce f by writing $y = f(x)$ or $x \rightsquigarrow f(x)$. If J is R or any closed interval, a function $f: J \to R$ may be regarded as determining a curve, which may be identified with its graph in $R \times R$.

Our single most important underlying assumption will be: in $\mathbb{S}$, *all curves determined by functions from R to R satisfy the Principle of Microstraightness.*

This assumption made, consider an arbitrary function $f: R \to R$. Since the curve $y = f(x)$ is microstraight around each of its points, there is a microsegment N of the curve $y = f(x)$ around the point (0, $f(0)$) which is straight, and so coincides with the tangent to the curve there. Now if f were a *polynomial* function, then N could be taken to be the image of Δ under f. To see this (Fig. 1.4) observe that if $f(x) = a_0 + a_1x + a_2x^2 + \cdots + a_nx^n$, then $f(\varepsilon) = a_0 + a_1\varepsilon$ for any ε in Δ, so that $(\varepsilon, f(\varepsilon))$ lies on the tangent to the curve at the point $(0, a_0)$. We shall assume that, in $\mathbb{S}$, this remains the case for an arbitrary function $f: R \to R$, in other words, that arbitrary functions from R to R behave locally like polynomials[20]. If we consider only the restriction g of f to Δ, this assumption entails that the graph of g is a piece of a unique straight line passing through the point (0,$g(0)$), in short, that g is *affine* on Δ. Thus we are led finally to suppose that the following basic postulate holds in $\mathbb{S}$, which we term the Principle of Microaffineness.

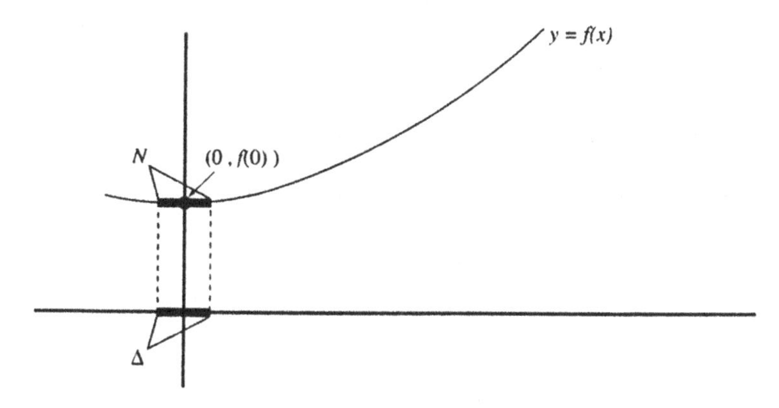

Fig. 1.4

Principle of Microaffineness *For any map $g: \Delta \to R$, there exists a unique b in R such that, for all ε in Δ, we have*

$$g(\varepsilon) = g(0) + b.\varepsilon.$$

This says that the graph of g is a straight line passing through (0, $g(0)$) with slope b.

The Principle of Microaffineness may be construed as asserting that, in $\mathbb{S}$, the microneighbourhood Δ can be subjected only to translations and rotations, i.e.

[20] The counterpart of this assumption in classical analysis is, of course, the fact that all smooth functions have Taylor expansions.

behaves as if it were an infinitesimal 'rigid rod'. Δ may also be thought of as a *generic tangent vector* because Microaffineness entails that it can be 'brought into coincidence' with the tangent to any curve at any point on it. Since we will shortly show that Δ does not reduce to a single point, it will be, so to speak, 'large enough' to have a slope but 'too small' to bend. Thus (as we have already remarked in the Introduction), Δ may be considered an entity possessing both location and direction, but lacking genuine extension, or in short, a pure synthesis of location and direction.

Let us assume that in $\mathbb{S}$ we can form the space R^{Δ} of all functions from Δ to R. If to each (a, b) in $R \times R$ we assign the function $\phi_{ab}: \Delta \to R$ defined by $\phi_{ab}(\varepsilon) = a + b\varepsilon$, it is easily seen that the Principle of Microaffineness is equivalent to the assertion that the resulting correspondence ϕ sending each (a,b) to ϕ_{ab} is a bijection between $R \times R$ and R^{Δ}.

Exercise

1.8 R^{Δ} is a ring with the natural operations $+$, . defined on it by $(f + g)(\varepsilon) = f(\varepsilon) + g(\varepsilon)$, $(f.g)(\varepsilon) = f(\varepsilon).g(\varepsilon)$ for f, g in R^{Δ}. Show that, if we define operations $\oplus$, $\odot$ on $R \times R$ by $(a, b) \oplus (c, d) = (a + c, b + d)$, $(a, b) \odot (c, d) = (ac, ad + bc)$, then $(R \times R, \otimes, \odot)$ is a ring and ϕ as defined above is a ring isomorphism.

We conclude this chapter by deriving some important properties of Δ.

Theorem 1.1 *In a smooth world* $\mathbb{S}$,

(i) Δ *is included in the closed interval* $[0, 0]$, *but is nondegenerate, i.e. not identical with* $\{0\}$.
(ii) *Every element of* Δ *is indistinguishable from* 0.
(iii) *It is false that, for all* ε *in* Δ, *either* $\varepsilon = 0$ *or* $\varepsilon \neq 0$.
(iv) Δ *satisfies the* Principle of (Universal) Microcancellation, *namely, for any* a, b *in* R, if $\varepsilon a = \varepsilon b$ *for all* ε *in* Δ, *then* $a = b$. *In particular, if* $\varepsilon a = 0$ *for all* ε *in* Δ, *then* $a = 0$.

Proof (i) That Δ is included in [0,0] follows immediately from exercise 1.6(i). Suppose that Δ did coincide with $\{0\}$. Consider the function $g: \Delta \to R$ defined by $g(\varepsilon) = \varepsilon$. Then $g(\varepsilon) = g(0) + b\varepsilon$ both for $b = 0$ and $b = 1$. Since $0 \neq 1$, this violates the uniqueness of b guaranteed by Microaffineness. Therefore Δ cannot coincide with $\{0\}$.

(ii) Suppose that, if possible, $\varepsilon^2 = 0$ and $\varepsilon \neq 0$. Then since R is a field, $1/\varepsilon$ exists and $\varepsilon.(1/\varepsilon) = 1$. Hence $0 = 0.(1/\varepsilon) = \varepsilon^2.(1/\varepsilon) = \varepsilon.(\varepsilon/\varepsilon) = \varepsilon.1 = \varepsilon$. Therefore the assumption $\varepsilon^2 = 0$, i.e. ε in Δ, is incompatible with the assumption that $\varepsilon \neq 0$. But this is the assertion made in (ii).

(iii) Suppose that

$*$ for any ε in Δ, either $\varepsilon = 0$ or $\varepsilon \neq 0$.

Then since by (ii) it is not the case that $\varepsilon \neq 0$, it follows that the first disjunct $\varepsilon = 0$ must hold for any ε. This, however, is in contradiction with (i). It follows that (*) must be false, which is (iii).

(iv) Suppose that, for all ε in Δ, $\varepsilon a = \varepsilon b$ and consider the function $g: \Delta \to R$ defined by $g(\varepsilon) = \varepsilon a$. The assumption then implies that g has both slope a and slope b: the uniqueness clause in Microaffineness yields $a = b$.

The proof is complete.

The Principle of Microcancellation should be carefully noted, since we shall be employing it constantly in order to ensure that microquantities (apart from 0) do not figure in the final results of our calculations. Observe that the cancellation of ε is only permissible when $\varepsilon a = \varepsilon b$ for all ε in Δ: it is of course not enough merely that $\varepsilon a = \varepsilon b$ for **some** ε in Δ. However, this latter possibility will not arise in practice because statements involving microquantities will invariably concern arbitrary, rather than particular, microquantities (with the exception, of course, of 0).

Exercises

1.9 Show that the following assertions are *false* in $\mathbb{S}$: (i) $\varepsilon.\eta = 0$ for all ε, η in Δ; (ii) Δ is microstable; (iii) $x^2 + y^2 = 0$ implies $x^2 = 0$ for every x, y in R. (Hint: use (iv) of Theorem 1.1.)

1.10 Call two points a, b in R *neighbours if* $(a - b)$ is in Δ. Show that the neighbour relation is reflexive and symmetric, but not transitive. (Hint: use the previous exercise.)

1.11 Show that any map $f: R \to R$ is *continuous* in the sense that it sends neighbouring points to neighbouring points. Use this to give another proof of part (ii) of Theorem 1.1.

1.12 Show that, for any $\varepsilon_1, \ldots, \varepsilon_n$ in Δ, we have $(\varepsilon_1 + \cdots + \varepsilon_n)^{n+1} = 0$.

1.13 Show that the following principle of Euclidean geometry is *false* in S:

Given any straight lines L, L' both passing through points p, p', either $p = p'$ or $L = L'$. (Hint: consider lines passing through the origin with slopes the microquantities ε, η respectively.)

But show that the following is *true* in *S*:

For any pair of *distinct* points, there is a unique line passing through them both.

2

Basic differential calculus

2.1 The derivative of a function

We turn next to the development of the differential calculus in a smooth world $\mathbb{S}$. We begin by defining the '*derivative*' of an arbitrary given function f: $R \to R$. For fixed x in R, define the function g_x: $\Delta \to R$ by

$$g_x(\varepsilon) = f(x + \varepsilon).$$

By Microaffineness there is a unique b in R, whose dependence on x we will indicate by denoting it b_x, such that, for all ε in Δ,

$$f(x + \varepsilon) = g_x(\varepsilon) = g_x(0) + b_x.\varepsilon = f(x) + b_x.\varepsilon. \tag{2.1}$$

Allowing x to vary then yields a function $x \rightsquigarrow b_x$: $R \to R$ which is written f' and called, as is customary, the *derivative* of f. If f is given as $y = f(x)$, we shall occasionally adopt the familiar notation dy/dx for f'. Equation (1), which may be written

$$f(x + \varepsilon) = f(x) + \varepsilon f'(x), \tag{2.2}$$

for arbitrary x in R and ε in Δ, is the *fundamental equation of the differential calculus* in $\mathbb{S}$. The quantity $f'(x)$ is the slope at x of the curve determined by f, and the microquantity

$$\varepsilon f'(x) = f(x + \varepsilon) - f(x)$$

is precisely the change or *increment* in the value of f on passing from x to $x + \varepsilon$[21].

[21] The exactness of the increment defined here is to be sharply contrasted with its 'approximate' counterpart in the classical differential calculus.

We see that in $\mathbb{S}$ every map $f\colon R \to R$ has a derivative. It follows that the process of forming derivatives can be iterated indefinitely[22] so as to yield higher derivatives f'', f''', ... Thus the *n*th derivative $f^{(n)}$ of f is defined recursively by the equation

$$f^{(n-1)}(x+\varepsilon) = f^{(n-1)}(x) + \varepsilon . f^{(n)}(x).$$

It should be clear that the definition of the derivative given above can be extended verbatim to any function defined on a microstable part of R, in particular, by exercise 1.7, on any closed interval.

In the remainder of this text we shall use the symbol J to denote an arbitrary closed interval or R itself.

Exercise

2.1 For ε, η, ζ in Δ, show that $f(x+\varepsilon+\eta) = f(x) + (\varepsilon+\eta)f'(x) + \varepsilon\eta f''(x)$ and $f(x+\varepsilon+\eta+\zeta) = f(x) + (\varepsilon+\eta+\zeta)f'(x) + (\varepsilon\eta + \varepsilon\zeta + \eta\zeta)f''(x) + \varepsilon\eta\zeta f'''(x)$. Generalize.

This definition of the derivative, together with the Principle of Microcancellation, enables the basic formulas of the differential calculus to be derived in a straightforward purely algebraic fashion. The proofs of some of the following examples are left as exercises to the reader.

Sum and scalar multiple rules For any functions f, $g\colon J \to R$ and any c in R,

$$(f+g)' = f' + g' \qquad (c.f)' = cf',$$

where $f+g$, $c.f$ are the functions $x \rightsquigarrow f(x) + g(x)$, $x \rightsquigarrow cf(x)$, respectively.

Product rule For any functions f, g $J \to R$, we have

$$(f.g)' = f'.g + f.g'$$

where $f.g$ is the function $x \rightsquigarrow f(x).g(x)$.

Proof We have, for any ε in Δ,

$$(f.g)(x+\varepsilon) = (f.g)(x) + \varepsilon.(f.g)'(x) = f(x).g(x) + \varepsilon.(f.g)'(x) \qquad (2.3)$$

[22] Thus in $\mathbb{S}$ every function f defined on R is smooth in the technical sense of possessing derivatives of all orders.

and

$$\begin{aligned} f(x+\varepsilon).g(x+\varepsilon) &= [f(x)+\varepsilon f'(x)].[g(x)+\varepsilon g'(x)] \\ &= f(x)g(x)+\varepsilon[f'(x)g(x)+f(x)g'(x)] \\ &\quad +\varepsilon^2.f(x)g(x). \end{aligned} \tag{2.4}$$

Equating (2.3) and (2.4) and recalling that $\varepsilon^2 = 0$ gives

$$\varepsilon(f.g)'(x) = \varepsilon.[f'(x)g(x) + f(x)g'(x)].$$

Since this is true for any ε, it may be cancelled to yield the desired result.

Polynomial rule If $f(x) = a_0 + a_1x + \cdots + a_nx^n$, then

$$f' = \sum_{k=1}^{n} ka_kx^{k-1}.$$

In particular $(cx)' = c$.

Quotient rule If g: $J \to R$ satisfies $g(x) \neq 0$ for all x in J, then for any f: $J \to R$

$$(f/g)' = (f'.g - f.g')/g^2,$$

where f/g is the function $x \rightsquigarrow f(x)/g(x)$.

Composite rule For any f, g: $J \to R$ we have

$$(g \circ f)' = (g' \circ f).f',$$

where $g \circ f$ is the function $x \rightsquigarrow g(f(x))$.

Proof We have

$$(g \circ f)(x+\varepsilon) = (g \circ f)(x) + \varepsilon(g \circ f)'(x) = g(f(x)) + \varepsilon(g \circ f)'(x). \tag{2.5}$$

Since $f(x+\varepsilon) = f(x) + \varepsilon f'(x)$ and (by exercise 1.6) $\varepsilon f'(x)$ is in Δ, it follows that

$$g(f(x+\varepsilon)) = g(f(x)+\varepsilon f'(x)) = g(f(x)) + \varepsilon f'(x).g'(f(x)). \tag{2.6}$$

Equating (2.5) and (2.6), removing common terms and finally cancelling ε yields the result.

Inverse Function rule Suppose that f: $J_1 \to J_2$ admits an *inverse*, that is, there exists a function g: $J_2 \to J_1$ such that $g(f(x)) = x$ and $f(g(y)) = y$ for all x in J_1, y in J_2. Then f' and g' are related by the equation

$$(f' \circ g).g' = (g' \circ f).f' = 1.$$

Proof By the polynomial rule the derivative of the function $i(x) = x$ is 1. So the result follows immediately from the Composite rule.

Observe that it follows from this last rule that the derivative of any function admitting an inverse cannot vanish anywhere.

2.2 Stationary points of functions

One of the most important applications of the differential calculus is in determining 'stationary points' of functions. This can be carried out in $\mathbb{S}$ by what we shall call the Method of Microvariations, which goes back in principle to Fermat.

We define a point a in R to be a *stationary point*, and $f(a)$ a *stationary value*, of a given function $f: R \to R$ if microvariations around a fail to change the value of f there, i.e. provided that, for all ε in Δ,

$$f(a+\varepsilon) = f(a).$$

Now this holds if and only if, for all ε,

$$f(a) + \varepsilon f'(a) = f(a+\varepsilon) = f(a),$$

i.e. if and only if $\varepsilon . f'(a) = 0$ for all ε, or $f'(a) = 0$ by microcancellation. This establishes the following rule.

Fermat's rule A point a is a stationary point of a function f if and only if $f'(a) = 0$.

It is instructive to examine in this connection an analysis actually carried out by Fermat (presented in detail on pp. 167–8 of Baron, 1969). He wishes to maximize the function $f(x) = x(b-x)$. He allows x to become $x+e$ and then puts, for a stationary value, $f(x)$ approximately equal to $f(x+e)$, i.e.

$$x(b-x) \approx (x+e)(b-x-e).$$

Then, removing common terms,

$$0 \approx be - 2xe - e^2,$$

so that, dividing by e,

$$0 \approx b - 2x - e.$$

To obtain the stationary value he now sets $e = 0$ (i.e. $f'(x) = 0$) and so obtains $2x = b$. Clearly if in this argument we replace the suggestive but somewhat vague notion of 'approximate equality' by literal equality, the process of 'dividing by e and setting $e = 0$' is tantamount to assuming $e^2 = 0$ and cancelling e. In other words, e is implicitly being treated as a nilsquare infinitesimal. It thus seems fair to claim that Fermat's method of determining stationary points is faithfully represented in $\mathbb{S}$ by the method of microvariations.

2.3 Areas under curves and the Constancy Principle

Another important traditional application of the calculus is in calculating the area under (or bounded by) a curve. Let us see how this can be effected in $\mathbb{S}$. Suppose given a function $f\colon J \to R$; for x in J let $A(x)$ be the area under the curve defined by the function $y = f(x)$ bounded by the x- and y-axes and the line with abscissa x parallel to the y-axis (see Fig. 2.1). The resulting function $A(x)$

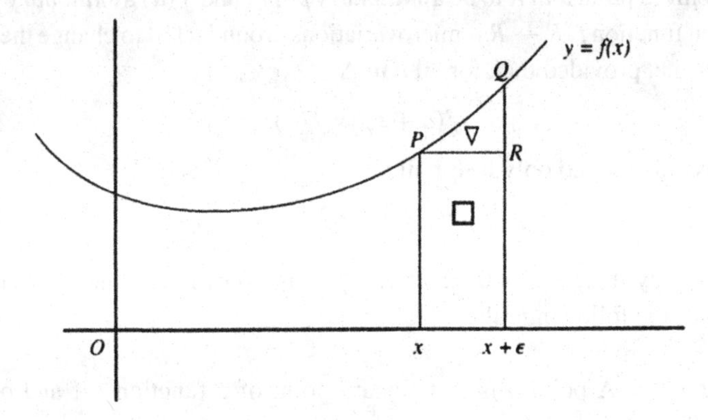

Fig. 2.1

will be called the *area function* associated with f: we are of course assuming that $A(x)$ is a well-defined function (this assumption will later be justified by the Integration Principle in Chapter 6). Then, by the Fundamental Equation, for ε in Δ,

$$A(x+\varepsilon) - A(x) = \varepsilon A'(x).$$

Now

$$A(x+\varepsilon) - A(x) = \square + \nabla,$$

where $\square$ is the area of the indicated rectangular region and ∇ is the area of PQR. Clearly $\square = \varepsilon f(x)$. Now, by Microstraightness, the microsegment PQ of the curve is straight, so ∇ is genuinely a triangle of base ε and height $f(x+\varepsilon) - f(x) = \varepsilon f'(x)$. Hence $\nabla = \left(\frac{1}{2}\right)\varepsilon.\varepsilon f'(x) = 0$ since $\varepsilon^2 = 0$. Therefore $\varepsilon A'(x) = A(x+\varepsilon) - A(x) = \square = \varepsilon f(x)$. Since this equality holds for arbitrary ε, we may cancel it on both sides to obtain[23] the following theorem.

[23] It will be apparent that our argument here is essentially a recapitulation of the discussion in the Introduction relating linear infinitesimals and indivisibles.

Fundamental Theorem of the Calculus *For any function f: $J \to R$, its area function satisfies*

$$A'(x) = f(x).$$

Now in order to be able to apply the Fundamental Theorem we need to assume that the following principle holds in $\mathbb{S}$:

Constancy Principle *If f: $J \to R$ is such that $f' = 0$ identically, then f is constant.*

It is easily shown that the Constancy Principle may be equivalently expressed in the form: if *f*: $J \to R$ satisfies $f(x + \varepsilon) = f(x)$ for all *x* in *J* and all ε in Δ – that is, if every point in the domain of *f* is a stationary point – then *f* is constant. It also follows immediately from the Constancy Principle that if two functions have identical derivatives, then they differ by at most a constant.

Exercise

2.2 Use the Fundamental Theorem and the Constancy Principle to deduce *Cavalieri's Principle* (1635): if two plane figures are included between a pair of parallel lines, and if the two segments cut by them on any line parallel to the including lines are in a fixed ratio, then the areas of the figures are in this same ratio.

The Constancy Principle, together with the Principle of Microaffineness, also has the following striking consequence. Call a part *U* of *R* *detachable*[24] if, for any *x* in *R*, it is the case that either *x* is in *U* or *x* is not in *U*. We can then prove the following theorem.

Theorem 2.1 *The only detachable parts of R are R itself and its empty part.*

Proof Suppose that *U* is a detachable part of *R*. Then the map *f* defined for *x* in *R* by

$$f(x) = 1 \text{ if } x \text{ is in } U,$$
$$f(x) = 0 \text{ if } x \text{ is not in } U,$$

is defined on the whole of *R*. We claim that the derivative f' of *f* is identically zero.

[24] The definition of detachability given here is easily seen to be equivalent to that of the Introduction. We call a part *U* of *R* 'detachable' because, if the condition is satisfied, *U* may be 'detached' from *R* leaving its 'complement' – the part of *R* consisting of all points not in *U* – cleanly behind.

To prove this, take any x in R and ε in Δ. Then

$$[f(x) = 0 \text{ or } f(x) = 1] \text{ and } [f(x+\varepsilon) = 0 \text{ or } f(x+\varepsilon) = 1].$$

Accordingly we have four possibilities:

(1) $f(x) = 0$ and $f(x+\varepsilon) = 0$
(2) $f(x) = 0$ and $f(x+\varepsilon) = 1$
(3) $f(x) = 1$ and $f(x+\varepsilon) = 0$
(4) $f(x) = 1$ and $f(x+\varepsilon) = 1$,

Since f is continuous and x, $x+\varepsilon$ are neighbouring points, cases (2) and (3) may be ruled out by exercise 1.11. This leaves cases (1) and (4), and in either of these we have $f(x) = f(x+\varepsilon)$. Hence for all x in R, ε in Δ,

$$\varepsilon f'(x) = f(x+\varepsilon) - f(x) = 0,$$

so that, cancelling ε, we obtain $f'(x) = 0$ for all x as claimed.

The Constancy Principle now implies that f is constant, that is, constantly 1 or constantly 0. In the former case, U is R, and in the latter, U is empty. The proof is complete.

Thus, in $\mathbb{S}$, the smooth line is *indecomposable* in the sense that it cannot be split in any way whatsoever into two disjoint nonempty parts. It is easy to extend this result to any closed interval.

2.4 The special functions

We shall need to introduce certain *special functions* into our smooth world $\mathbb{S}$. The first of these is the *square root* function $\sqrt{x}$ or $x^{\frac{1}{2}}$. We regard this function as being defined on, and taking values in, R^+, the part of R consisting of all x for which $x > 0$. By exercise 1.6(iv), R^+ is microstable, so $x^{\frac{1}{2}}$ has a derivative, also with domain and codomain R^+. Using the inverse function rule, it is easy to show that this derivative is the function

$$x \to \left(\tfrac{1}{2}\right) 1/\sqrt{x} = \left(\tfrac{1}{2}\right) x^{-\frac{1}{2}},$$

We shall also need to assume the presence in $\mathbb{S}$ of the familiar *sine* and *cosine* functions sin: $R \to R$ and cos: $R \to R$. As usual, if a, b, c are the sides of a right-angled triangle with base angle x (measured in radians), we have

$$a = c\cos x \qquad b = c\sin x,$$

We assume the familiar relations

$$\sin 0 = 0 \qquad \cos 0 = 1$$
$$\sin^2 x + \cos^2 x = 1$$
$$\sin(x + y) = \sin x \cos y + \cos x \sin y$$
$$\cos(x + y) = \cos x \cos y - \sin x \sin y.$$

We now determine $\sin \varepsilon$ and $\cos \varepsilon$ for ε in Δ. Consider a segment of angle 2ε (radians) of a circle of unit radius (Fig. 2.2). By Microstraightness, the arc AB

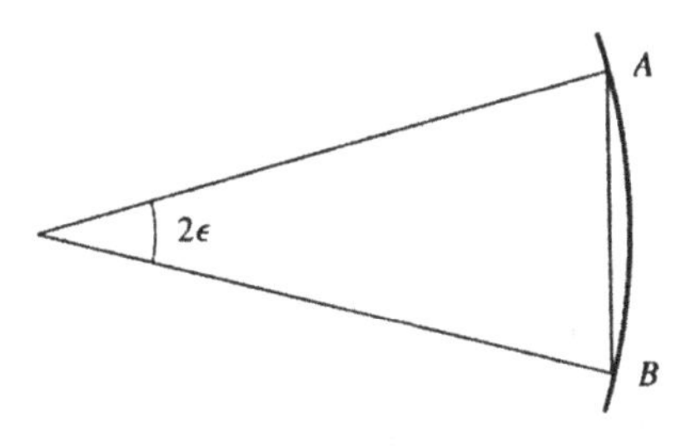

Fig. 2.2

of length 2ε is straight and therefore coincides with the chord AB, which has length $2 \sin \varepsilon$. Accordingly $2\varepsilon = 2 \sin \varepsilon$, so that

$$\sin \varepsilon = \varepsilon.$$

Moreover,

$$1 = \sin^2 \varepsilon + \cos^2 \varepsilon = \varepsilon^2 + \cos^2 \varepsilon = \cos^2 \varepsilon,$$

so that

$$\cos \varepsilon = 1.$$

These facts enable us to determine the derivatives of sin and cos. For ε in Δ we have

$$\sin x + \varepsilon \sin' x = \sin (x + \varepsilon) = \sin x \cos \varepsilon + \cos x \sin \varepsilon = \sin x + \varepsilon \cos x,$$

so that $\varepsilon \sin' x = \varepsilon \cos x$. Therefore, cancelling ε,

$$\sin' x = \cos x.$$

A similar calculation gives

$$\cos' x = - \sin x.$$

Let $f: J \to R$ be any function. At a point A with abscissa x on the curve determined by f let $\phi(x)$ be the angle the tangent to the curve there makes with the horizontal x-axis (see Fig. 2.3). For ε in Δ, let B the point on the curve

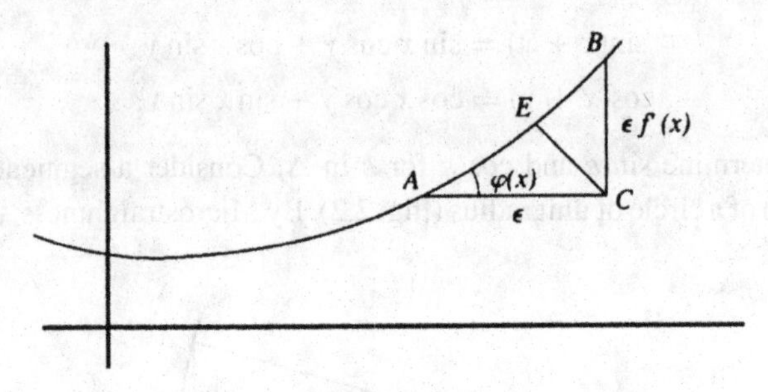

Fig. 2.3

with abscissa $x + \varepsilon$. Then by Microstraightness the arc AB of the curve is straight and so we may consider the microtriangle ABC as indicated in Fig. 2.3, where C is the point $(x + \varepsilon, f(x))$. Clearly AB, BC have lengths $\varepsilon, \varepsilon f'(x)$, respectively. Now let E be the foot of the perpendicular from C to the line AB. Then, indicating lengths of lines by underscoring[25], we have

$$\underline{AC}.\sin\phi(x) = \underline{CE} = \underline{BC}.\cos\phi(x),$$

i.e.

$$\varepsilon \sin\phi(x) = \varepsilon f'(x)\cos\phi(x).$$

Cancelling ε gives the fundamental relation[26]

$$\sin\phi(x) = f'(x)\cos\phi(x). \tag{2.7}$$

It follows from this that

$$1 - \cos^2\phi(x) = \sin^2\phi(x) = f'(x)^2\cos^2\phi(x),$$

from which we infer

$$\cos\phi(x) = 1/\surd(1 + f'(x)^2).$$

In particular, $\cos\phi(x)$ must always be $\neq 0$.

The other special function we shall consider to be present in $\mathbb{S}$ is the *exponential* function. For our purposes this will be characterized as a function $h: R \to R$

[25] In the remainder of this text we shall employ this device without comment.

[26] In classical analysis this equation would be expressed as $f'(x) = \tan\theta(x)$, where tan is the usual tangent function. However, since tan is not defined on the whole of R (being undefined at $(2n + 1)\pi/2$ for integral n), we prefer to avoid its use in $\mathbb{S}$.

possessing the two following properties: $h(x) > 0$ for all x in R; $h' = h$. Observe that these conditions determine h up to a multiplicative constant. For if g is another function satisfying them, we may consider the derivative of the function g/h:

$$(g/h)' = (g'h - h'g)/h^2 = (gh - hg)/h^2 = 0.$$

Therefore, by the Constancy Principle, $g/h = c$ for some c in R, whence $g = c.h$. In particular, if $g(0) = h(0)$, then $c = 1$ and $g = h$.

We may suppose without loss of generality that $h(0) = 1$, since, if necessary, h may be replaced by the function $h/h(0)$. Under these conditions we write exp for h and call it the *exponential function.* The function exp: $R \to R$ is thus characterized uniquely by the following conditions:

$$\exp(x) > 0 \qquad \exp' = \exp \qquad \exp(0) = 1.$$

Exercise

2.3 Show that, for any a, b in R, $b.\exp(ax)$ is the unique function u: $R \to R$ satisfying the conditions $u(x) > 0$, $u(0) = b$, $u' = au$.

Note that, for ε in Δ,

$$\exp(\varepsilon) = \exp(0) + \varepsilon \exp'(0) = 1 + \varepsilon.$$

Note also that exp satisfies the equation

$$\exp(x + y) = \exp(x).\exp(y). \tag{2.8}$$

For let y have a fixed but arbitrary value a. Then

$$[(\exp(x + a)/\exp(x)]' = [\exp(x)\exp'(x + a) \\ - \exp'(x)\exp(x + a)]/\exp(x)^2 = 0.$$

So, by the Constancy Principle, there is b in R such that, for all x in R

$$\exp(x + a)/\exp(x) = b.$$

Taking $x = 0$, we have

$$b = \exp(a)/\exp(0) = \exp(a).$$

Therefore, for arbitrary a in R,

$$\exp(x + a) = b \exp(x) = \exp(x)\exp(a),$$

as claimed.

Exercise

2.4 Show that $\exp(-x) = 1/\exp(x)$.

If, as usual, we write e for $\exp(1)$, it follows from (2.8) that, for any natural number n,

$$\exp(n) = \exp(1).\exp(1).\ldots.\exp(1)(n \text{ times}) = e^n.$$

Thus $1 = \exp(0) = \exp(n - n) = \exp(n)\exp(-n)$, so that

$$\exp(-n) = 1/\exp(n) = e^{-n}.$$

For any rational number m/n we then have

$$(\exp(m/n))^n = \exp((m/n).n) = \exp(m) = e^m,$$

so that

$$\exp(m/n) = e^{m/n}.$$

Thus, for rational values of x, $\exp(x)$ behaves like the power e^x.

3
First applications of the differential calculus

In this chapter we turn to some of the traditional applications of the calculus, namely, the determination of areas, volumes, arc lengths, and centres of curvature. The arguments here take the form of direct computations, based on the analysis of figures: they can be rigorized quite easily by introducing the definite integral function over a closed interval, which will be deferred until Chapter 6.

In these applications, as well as in the physical applications to be presented in subsequent chapters, the role of infinitesimals will be seen to be twofold. First, as straight microsegments of curves, they play a 'geometric' role, enabling each infinitesimal figure to be taken as rectilinear, and as a result, ensuring that its area or volume, as the case may be, is a definite calculable quantity. And second, as nilsquare quantities, they play an 'algebraic' role in reducing the results of these calculations to a simple form, from which the desired result can be obtained by the Principle of Microcancellation[27].

3.1 Areas and volumes

We begin by determining the area of a circle. The method here – which was employed by Kepler (see Baron 1969) – is to consider the circle as being composed of a plurality of small isoceles triangles, each with its base on the circumference and apex at the centre (Fig. 3.1). Thus let $C(x)$ be the area of the sector OPQ of the given circle, where Q has abscissa x. Let $s(x)$ be the length of the arc PQ. Let R be the point on the circle with abscissa $x + \varepsilon$, where ε is in Δ. Then we have

$$\varepsilon C'(x) = C(x + \varepsilon) - C(x) = \text{ area } OQR.$$

[27] In this way we are, in the words of Weyl (1922, p. 92), employing 'the principle of gaining knowledge of the external world from the behaviour of its infinitesimal parts'.

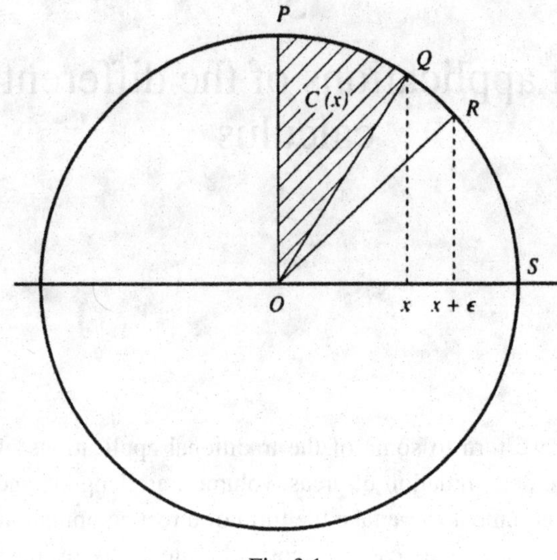

Fig. 3.1

Now, by Microstraightness, QR is a straight line of length

$$s(x+\varepsilon) - s(x) = \varepsilon s'(x),$$

so that, writing r for the circle's radius, OQR is a triangle of area

$$\left(\tfrac{1}{2}\right)r.\underline{QR} = \left(\tfrac{1}{2}\right)\varepsilon r s'(x).$$

Therefore

$$\varepsilon C'(x) = \left(\tfrac{1}{2}\right)\varepsilon r s'(x),$$

so that, cancelling ε on both sides,

$$C'(x) = \left(\tfrac{1}{2}\right)r s'(x).$$

Since $C(0) = s(0) = 0$, the Constancy Principle now yields

$$C(x) = \left(\tfrac{1}{2}\right)r s(x).$$

Hence, taking $x = r$,

$$\text{area of quadrant } OPS = \left(\tfrac{1}{2}\right)r.\underline{PS},$$

so that, multiplying both sides by 4,

$$\text{area of circle} = \left(\tfrac{1}{2}\right)r.\text{circumference}.$$

Assuming that the circumference of a circle is proportional to its radius (by the customary factor 2π), we thus arrive at the familiar formula for the area of a circle:

$$A = \pi r^2.$$

Exercises

3.1 Using a method similar to that just employed for determining the area of a circle, show that the area of the curved surface of a cone is πrh, where r is its base radius and h is the height of its curved surface. Deduce that the area of the curved surface of a frustum of a cone is $\pi(r_l + r_2)h$, where r_1 and r_2 are its top and bottom radii and h is the height of its curved surface.

3.2 Use the formula for the area of a circle and Cavalieri's Principle (exercise 2.2) to show that the area of an ellipse with semiaxes of lengths a, b is πab.

We next determine the volume of a cone. (The germ of the argument here seems to have origininated with Democritus.) Referring to Fig. 3.2, let $V(x)$ be

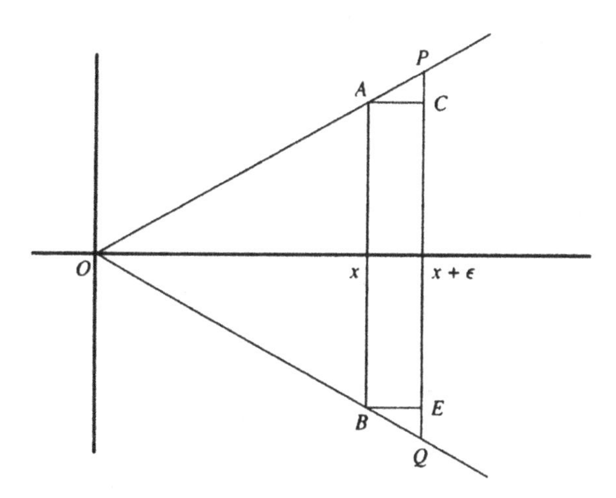

Fig. 3.2

the volume of the section OAB of the cone of length x, where OA has slope b. Then for ε in Δ, we have

$$\begin{aligned} \varepsilon V'(x) &= V(x+\varepsilon) - V(x) \\ &= \text{volume of } APQB \text{ rotated about } x\text{-axis} \\ &= \text{volume of } ACEB \text{ rotated about } x\text{-axis} + \text{volume of } ACP \text{ rotated about } x\text{-axis}. \end{aligned}$$

Now since the area of ACP is $\left(\frac{1}{2}\right)\varepsilon.b\varepsilon = 0$, it follows that the volume of any figure obtained by rotating it is also zero. Therefore, using the formula for the area of a circle,

$$\varepsilon V'(x) = \text{volume of } ACEB \text{ rotated about } x\text{-axis} = \varepsilon\pi b^2x^2.$$

Cancelling ε on both sides gives

$$V'(x) = \pi b^2x^2. \tag{3.1}$$

It now follows from the polynomial rule and the Constancy Principle that

$$V(x) = \left(\tfrac{1}{3}\right)\pi b^2x^3 + k,$$

where k is a constant. Since $V(0) = 0$, $k = 0$ and so we get[28]

$$V(x) = \left(\tfrac{1}{3}\right)\pi b^2x^3. \tag{3.2}$$

Thus if h is the cone's height we obtain finally

$$\text{volume of cone} = V(h) = \left(\tfrac{1}{3}\right)\pi b^2h^3 = \left(\tfrac{1}{3}\right)\pi h(bh)^2 = \left(\tfrac{1}{3}\right)\pi r^2h.$$

Exercise

3.3 Show that the volume of a conical frustum of top and bottom radii r_1 and r_2 and altitude h is $\left(\frac{1}{3}\right)\pi h\left(r_1^2 + r_1r_2 + r_2^2\right)$.

The formula for a volume of a cone figures in the ingenious method – due to Archimedes – that we shall employ in $\mathbb{S}$ for determining the volume of a sphere. This method uses the concept of the *moment* of a body about a point (or line) which is defined to be the product of the mass of the body and the distance of its centre of gravity from the point (or line).

Consider a sphere of radius r, positioned with its polar diameter along the x-axis with its north pole N at the origin (Fig. 3.3: here we see a circular cross-section of the sphere cut off by a plane passing through its centre). By rotating the rectangle $NABS$ and the triangle NCS, a cylinder and a cone are obtained. We assume that the sphere, cone and cylinder are homogeneous solids of unit density, so that the mass of each, and that of any part thereof, concides with its volume.

Let T be a point on the x-axis at distance r from the origin in the opposite direction from S. For θ in $[0, \pi/2]$ consider now the line passing through N at angle θ with NA, intersecting the circle at P. The line through P parallel to BS

[28] In drawing similar conclusions in the remainder of this text we shall omit reference to the polynomial rule, the Constancy Principle and the constant k (when this latter turns out to be zero). Thus we shall merely say, for example, that an equation like (3.1) yields the corresponding equation (3.2), without further comment.

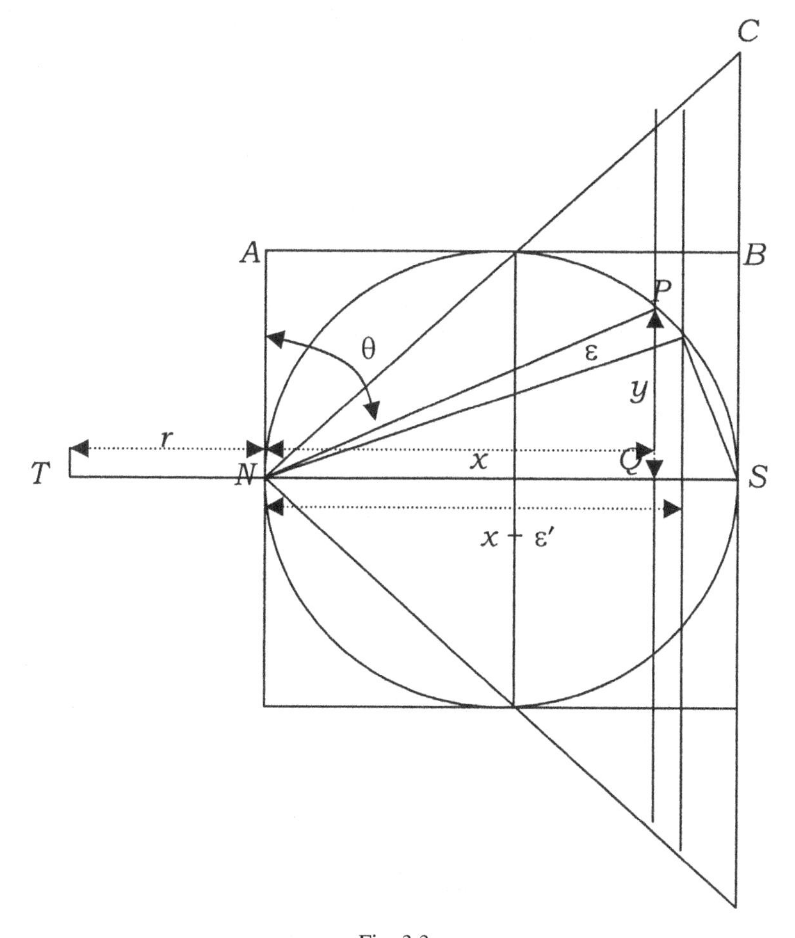

Fig. 3.3

intersects TS at a point Q at distance $x = x(\theta)$ from N and $y = y(\theta)$ from P. By elementary trigonometry, $x = 2r\sin^2\theta$ and $y = 2r\sin\theta\cos\theta$.

For $i = 1, 2, 3$ let $V_i(\theta)$ be the volume (= mass) of the segment of the sphere, cone and cylinder, respectively, cut off at a distance x from N. Also for $i = 1$, 2 let

$$M_i(\theta) = \text{moment about } N \text{ of the whole mass of } V_i(\theta) \text{ concentrated at } T$$

and

$$M_3(\theta) = \text{moment about } N \text{ of the whole mass of } V_i(\theta) \text{ left where it is.}$$

Now allow θ to vary to $\theta + \varepsilon$, with ε in Δ. Then x varies to $x + \varepsilon'$ with

$$\varepsilon' = x(\theta + \varepsilon) - x(\theta) = 2r[\sin^2(\theta + \varepsilon) - \sin^2\theta] = 4\varepsilon r\sin^2\theta\cos\theta.$$

By a calculation similar to that performed for the volume of a cone, we find that

$$\varepsilon M_1'(\theta) = \varepsilon' r.\pi y^2 = 4\varepsilon\pi r^3 \sin^2\theta\cos^2\theta = 16\varepsilon\pi r^4 \sin^3\theta\cos^3\theta$$
$$\varepsilon M_2'(\theta) = \varepsilon' r.\pi x^2 = 4\varepsilon'\pi r^3 \sin^4\theta = 16\varepsilon\pi r^4 \sin^5\theta\cos\theta$$
$$\varepsilon M_3'(\theta) = \varepsilon' x.\pi r^2 = \varepsilon'\pi r^3 \sin^2\theta = 8\varepsilon\pi r^4 \sin^3\theta\cos\theta.$$

It follows that

$$\varepsilon[M_1'(\theta) + M_2'(\theta)] = 16\varepsilon\pi r^4 \sin^3\theta\cos\theta[\cos^2\theta + \sin^2\theta]$$
$$= 16\varepsilon\pi r^4 \sin^3\theta\cos\theta = 2\varepsilon M_3'(\theta);$$

cancelling ε on both sides gives

$$M_1'(\theta) + M_2'(\theta) = 2M_3'(\theta).$$

So if in this equation we set $\theta = \pi/2$, we get

(∗) moment of (mass of sphere + mass of cone concentrated at T) about N $= 2 \times$ moment of mass of cylinder about N.

Write $V_i = V_i(\pi/2), i = 1, 2, 3$ for the volume of the sphere, cone and cylinder, respectively. Then the left-hand side of (*) is $r(V_1 + V_2)$ and the right-hand side is $2rV_3$ (since, by symmetry, the centre of mass of the cylinder coincides with its geomertrical centre). Equating these and cancelling r gives

$$V_1 + V_2 = 2V_3.$$

Using the formula already obtained for the volume of a cone and the evident formula for the volume of a cylinder, we obtain from this last equation

$$V_1 + 8\pi r^3/3 = 4\pi r^3,$$

giving finally

$$V_1 = 4\pi r^3/3.$$

Remark The volume of a sphere can also be calculated in S by more conventional means: we leave this as an exercise of the reader.

3.2 Volumes of revolution

In Fig. 3.4, suppose that the curve AB with equation $y = f(x)$ makes a complete revolution about the x-axis OX, thereby tracing out a surface. The section

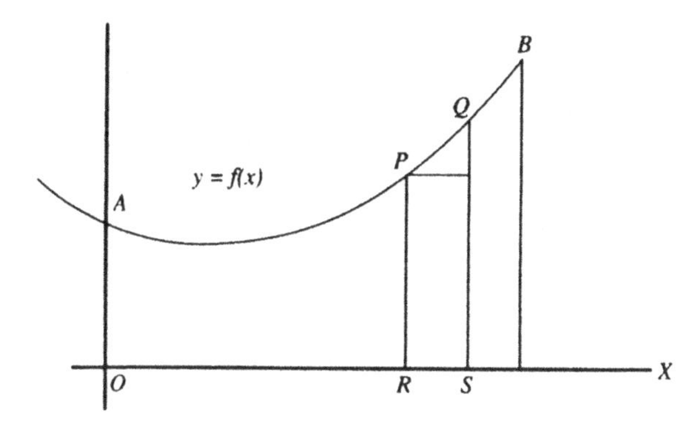

Fig. 3.4

of the surface by any plane perpendicular to OX – the axis of revolution – is a circle. We wish to find the volume intercepted between the surface and planes perpendicular to OX passing through A and B.

Write $V(x)$ for the volume intercepted between the planes through O and P perpendicular to OX, where P has abscissa x. Let Q be a point on the curve with abscissa $x + \varepsilon$, with ε in Δ. Then by Microstraightness the arc PQ is straight and we have

$$\varepsilon V'(x) = V(x+\varepsilon) - V(x)$$

$$= \text{volume of conical frustum obtained by rotating } PRSQ \text{ about } OX.$$

By exercise 3.3, this last quantity is

$$\begin{aligned}
&\left(\tfrac{1}{3}\right) \ \pi.\underline{RS}(\underline{PR}^2 + \underline{PR}.\underline{QS} + \underline{QS}^2)\\
&= \left(\tfrac{1}{3}\right)\pi[f(x)^2 + f(x)f(x+\varepsilon) + f(x+\varepsilon)^2]\\
&= \left(\tfrac{1}{3}\right)\pi\varepsilon[f(x)^2 + f(x)(f(x)+\varepsilon f'(x)) + f(x)^2 + 2\varepsilon f(x) + \varepsilon^2 f'(x)^2]\\
&= \varepsilon\pi f(x)^2,
\end{aligned}$$

using $\varepsilon^2 = 0$. Cancelling ε on both sides of the equation gives the relation

$$V'(x) = \pi f(x)^2.$$

As an example we calculate the volume of a torus. A *torus* or *anchor ring* is the surface generated by a circle which revolves about an axis in its plane, the axis not intersecting the circle (although it may be tangent to it). Let r be the radius of the circle and c the distance of its centre from the axis (Fig. 3.5). The equation of the circle may then be taken to be $x^2 + (y-c)^2 = r^2$.

Let B_1, B_2 be the points of intersection with the circle of a line drawn parallel to OP_1 in such a way that the area of the segment $P_1P_2B_2B_1$ is exactly half the

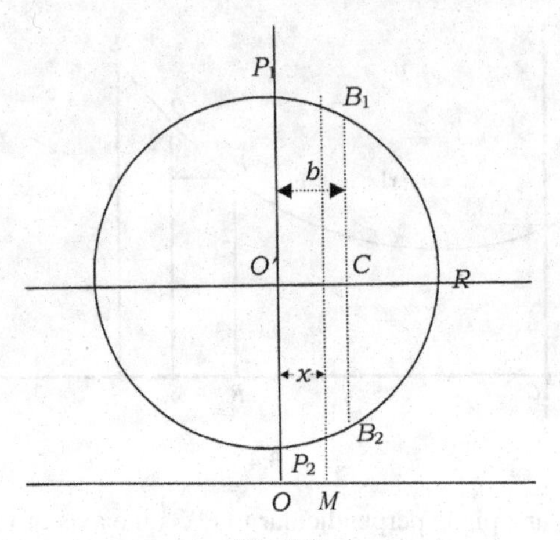

Fig. 3.5

area of the semicircle P_1RP_2. Write b for the length of $O'C$ where C is the point of intersection of the B_1B_2 with $O'R$; clearly $b < r$.

Now the equation of the arc P_1B_1 is

$$y = f_1(x) = c + (r^2 - x^2)^{\frac{1}{2}}$$

and that of P_2B_2 is

$$y = f_2(x) = c - (r^2 - x^2)^{\frac{1}{2}}$$

for x in $[0, b]$[29].

For x in $[0, b]$ let $V(x)$ be the volume of the torus intercepted between planes prependicular to the x-axis passing through O and M where M is at distance x from O. Let $V_1(x)$, $V_2(x)$ be the volumes similarly intercepted of the surfaces of revolution swept out by the curves P_1B_1 and P_2B_2. Then $V(x) = V_1(x) - V_2(x)$, so that

$$V'(x) = V_1{}'(x) - V_2{}'(x).$$

By the formula for the volume of revolution obtained above, we have

$$V_1{}'(x) = \pi f_1(x)^2 = \pi\left[c + (r^2 - x^2)^{\frac{1}{2}}\right]^2$$
$$V_2{}'(x) = \pi f_2(x)^2 = \pi\left[c - (r^2 - x^2)^{\frac{1}{2}}\right]^2.$$

Accordingly

$$V_1{}'(x) = \pi\left[\left(c + (r^2 - x^2)^{\frac{1}{2}}\right)^2 - \left(c - (r^2 - x^2)^{\frac{1}{2}}\right)^2\right] = 4\pi c(r^2 - x^2)^{\frac{1}{2}}. \quad (3.3)$$

[29] The point here is that the square root function in $\mathbb{S}$ is defined only for strictly positive arguments so that f_1 and f_2 cannot be regarded as being defined on $[0, r]$. However, since $b < r$, f_1 f_2 *are* legitimate functions on $[0, b]$. This subtlety was overlooked in the first edition of this book.

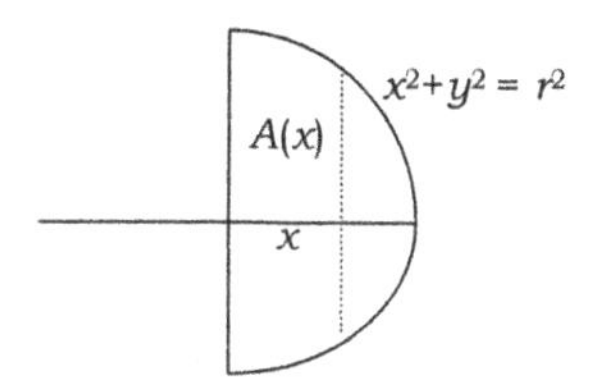

Now for x in $[O, b]$, write $A(x)$ for the area of the circle $x^2 + y^2 = r^2$ intercepted by the y-axis and a line parrallel to it at a distance x from the circle's centre (see figure immediately above). By the Fundamental Theorem of Calculus,

$$\frac{1}{2}A'(x) = y = (r^2 - x^2)^{\frac{1}{2}}.$$

Hence, by (3.3), $V'(x) = 2\pi c A'(x)$, whence $V(x) = 2\pi c A(x)$. Therefore, since $A(b)$ is half the area of the semicircle,

$$V(b) = 2\pi c A(b) = 2\pi c.\frac{\pi r^2}{4} = \frac{1}{2}\pi^2 r^2 c.$$

By symmetry, $V(b)$ is one-fourth the volume of the torus, so it follows that the volume of the torus is

$$4V(b) = 2\pi^2 r^2 c.$$

Exercises

3.4 A *prolate* (*oblate*) *spheroid* is the surface generated by an ellipse which revolves around its major (minor) axis. If the major and minor axes are of lengths $2a$ and $2b$, respectively, show that the volume of the prolate spheroid is $4\pi ab^2/3$ and the oblate $4\pi a^2 b/3$.

3.5 Show that the volume of a spherical cap of height h is $\pi h^2[r - (\frac{1}{3})h]$, where r is the radius of the sphere.

3.6 Show that the volume intercepted between the plane through $x = h$ perpendicular to the x-axis and the paraboloid generated by the revolution of the parabola $y^2 = 4ax$ about the x-axis is $2\pi ah^2$.

3.3 Arc length; surfaces of revolution; curvature

We next show how to derive the formula for the arc length of a curve. Let $s(x)$ be the length of the curve C with equation $y = f(x)$ measured from a prescribed point O on it (Fig. 3.6). Given a point P on C with abscissa x, consider a neighbouring point Q on C with abscissa $x + \varepsilon$, where ε is in Δ. Let $\phi(x)$ be the angle

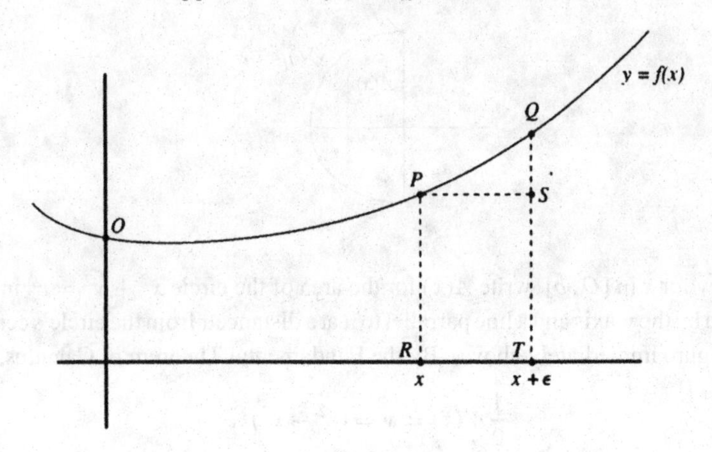

Fig. 3.6

that the tangent to the curve at P makes with the x-axis. Then, by Microstraightness, PQ is a straight line of length $s(x+\varepsilon) - s(x) = \varepsilon s'(x)$, and we have

$$\underline{PQ}.(1 + f'(x)^2)^{-\frac{1}{2}} = \underline{PQ}.\cos\phi(x) = \underline{PS} = \varepsilon.$$

Hence

$$\varepsilon s'(x) = \underline{PQ} = \varepsilon(1 + f'(x)^2)^{\frac{1}{2}}.$$

Cancelling ε yields the familiar equation

$$s'(x) = (1 + f'(x)^2)^{\frac{1}{2}}. \tag{3.4}$$

If the curve is defined parametrically by the equations

$$x = x(t) \qquad y = y(t),$$

then s may be regarded as a function $s(t)$ of t, and we have

$$s'(t) = s'(x)x'(t) \qquad y'(t) = y'(x)x'(t) = f'(x)x'(t).$$

Squaring (3.4) and multiplying by $x'(t)^2$ then gives

$$\begin{aligned} s'(t)^2 = s'(x)^2x'(t)^2 = x'(t)^2[1 + f'(x)^2] &= x'(t)^2 + x'(t)^2 f'(x)^2 \\ &= x'(t)^2 + y'(t)^2, \end{aligned}$$

so that

$$s'(t) = [x'(t)^2 + y'(t)^2]^{\frac{1}{2}}.$$

The equation for arc length is used in calculating areas of surfaces of revolution. Thus let $S(x)$ be the area of the surface traced out by the revolution of the arc OP about the x-axis (Fig. 3.6). Since the arc PQ is straight, we have

$$\begin{aligned}\varepsilon S'(x) &= S(x+\varepsilon) - S(x)\\ &= \text{area of surface of conical frustum traced by rotating of } PQ\\ &= \pi(\underline{RP} + \underline{TQ}).\underline{PQ} \text{ (by exercise 3.1)}\\ &= \pi[f(x) + f(x+\varepsilon)][s(x+\varepsilon) - s(x)]\\ &= \pi[2f(x) + \varepsilon f'(x)].\varepsilon s'(x) = 2\varepsilon\pi f(x)s'(x).\end{aligned}$$

Hence, cancelling ε and using (3.4) above,

$$S'(x) = 2\pi f(x)s'(x) = 2\pi f(x)[1 + f'(x)^2]^{\frac{1}{2}}. \tag{3.5}$$

Exercise

3.7 Use formula (3.5) to show that the area of a spherical cap of height h is $2\pi rh$, where r is the radius of the sphere.

We define the *curvature* of the curve $y = f(x)$ at P (Fig. 3.6) to be the 'rate of change' of $\phi(x)$ with respect to microvariations in arc length, i.e. it is that quantity $\kappa = \kappa(x)$ such that, for all ε in Δ,

$$\kappa.\underline{PQ} = \phi(x+\varepsilon) = \phi(x) = \varepsilon\phi'(x). \tag{3.6}$$

In order to derive an explicit formula for κ, we start with the fundamental relation (2.7)

$$\sin\phi(x) = f'(x)\cos\phi(x).$$

If we now form the derivatives of both sides of this equation we obtain

$$\begin{aligned}\phi'(x)\cos\phi(x) &= f''(x)\cos\phi(x) - \phi'(x)f'(x)\sin\phi(x)\\ &= f''(x)\cos\phi(x) - \phi'(x)f(x)^2\cos\phi(x).\end{aligned}$$

Since $\cos\phi(x) \neq 0$, it may be cancelled on both sides of this equation to yield

$$\phi'(x) = f''(x) - f'(x)^2\phi'(x),$$

whence

$$\phi'(x) = f''(x)/(1 + f'(x)^2). \tag{3.7}$$

Now

$$\underline{PQ} = \varepsilon s'(x) = \varepsilon(1 + f'(x)^2)^{\frac{1}{2}}$$

by (3.4). Substituting into (3.6) this expression for $\underline{PQ}$ and the expression for $\phi'(x)$ given by (3.7), we obtain

$$\varepsilon\kappa(1 + f'(x)^2)^{\frac{1}{2}} = \varepsilon f''(x)/(1 + f'(x)^2).$$

Since this holds for any ε in Δ, we may cancel it on both sides and thereby arrive at the well-known expression for the curvature of a curve at a point:

$$\kappa(x) = f''(x)/(1 + f'(x)^2)^{3/2}.$$

We now determine the location of the centre of curvature at a point on the curve. In $\mathbb{S}$, this may be done by the traditional method of intersection of consecutive normals (a device employed by Newton: see Chapter 7 of Baron, 1969). Thus let P and Q be points on the curve $y = f(x)$ with abscissae $x_0, x_0 + \varepsilon$ and let N_P be the normal to the curve at P. The normal N_Q to the curve at Q is called a consecutive normal from N. The *centre of curvature* of the curve at P is defined to be the common point of intersection of all consecutive normals from N. The coordinates of this point are easily determined as follows. Write $y_0 = f(x_0)$, $f_0{}' = f'(x_0)$, $f_0{}'' = f''(x_0)$. The equation of N_P is

$$(y - y_0)f_0{}' + x - x_0 = 0 \tag{3.8}$$

and that of N_Q is

$$[y - f(x_0 + \varepsilon)]f'(x_0 + \varepsilon) + x - x_0 - \varepsilon = 0,$$

i.e.

$$(y - y_0 - \varepsilon f_0{}') + \varepsilon(y - y_0)f_0{}'' + x - x_0 - \varepsilon = 0. \tag{3.9}$$

The coordinates (x,y) of the centre of curvature at P are the pair of solutions, for all ε in Δ, of this pair of equations. Substituting in (3.9) the value of $x - x_0$ given by (3.8) yields

$$\varepsilon(1 + f_0{}'^2) = \varepsilon f_0{}''.(y - y_0).$$

Since this equation is to hold for all ε, we may cancel ε on both sides to obtain

$$1 + f_0{}'^2 = f_0{}''.(y - y_0)$$

whence

$$y = y_0 + (1 + f_0{}'^2)/f_0{}''$$

and

$$x = x_0 - f_0{}'(1 + f_0{}'^2)/f_0{}''.$$

Exercise

3.8 (i) The distance between the centre of curvature at a point on a curve and the point is called the *radius of curvature* of the curve at the point. Show that, with the above notation, the radius of curvature at a point on $y = f(x)$ with abscissa x_0 is the reciprocal of the curvature there, that is, $(1 + f_0'^2)^{3/2}/f_0''$.

(ii) Let P be a point on a curve, let C be the centre of curvature and r the radius of curvature of the curve at P. The circle of radius r and centre C is called the *circle of curvature* (or *osculating circle*) of the curve at P. Show that the circle of curvature has the same curvature, and the same tangent at P, as the curve.

The curvature of lines in $\mathbb{S}$ is the source of a curious geometric phenomenon with whose description we conclude this chapter. On the curve with equation $y = f(x)$, consider neighbouring points P, Q with abscissae $x_0, x_0 + \varepsilon$ (Fig. 3.7). Moving the origin of coordinates to P transforms the variables x, y to u, v given

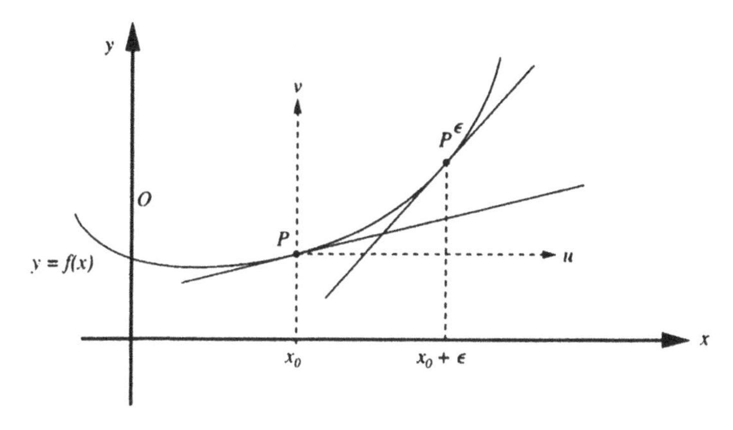

Fig. 3.7

by $u = x - x_0$, $v = y - f(x_0)$. Writing $f'(x_0) = a$, $f''(x_0) = b$, the equation of the tangent to the curve at P is, in terms of the new coordinates,

$$v = au, \tag{3.10}$$

and that of the tangent at Q

$$v = (a + b\varepsilon)u. \tag{3.11}$$

It is readily checked that both of these lines pass through the points P and Q, but (assuming $b \neq 0$), we cannot affirm their identity since we cannot do so for their slopes $a, a + b\varepsilon$. That is, although in $\mathbb{S}$ any two distinct points determine a unique line, two neighbouring points do not necessarily do so[30].

If the variable u just introduced is restricted to microvalues (i.e. values in Δ), then the resulting line segments (3.10) and (3.11) represent the straight microsegments of the curve around P and Q, respectively. In passing from P to Q the straight microsegment is subjected to a microincrease in slope of the amount $b\varepsilon$. That is, the 'curvature' of a curve in $\mathbb{S}$ is manifested in the microrotation of its straight microsegment as one moves along it.

[30] In fact, two neighbouring points need not determine any line at all. For example, if ε,η are members of Δ which are not proportional in the sense to be defined in the next chapter, there is no straight line passing through the points (0, 0) and (ε, η).

4
Applications to physics

In this chapter we present some of the traditional applications of the differential calculus to physical problems.

4.1 Moments of inertia

Suppose that we are given a (flat) surface of uniform density ρ (Fig. 4.1). Consider a rectangular microelement E of the surface with sides of lengths ε, η in Δ,

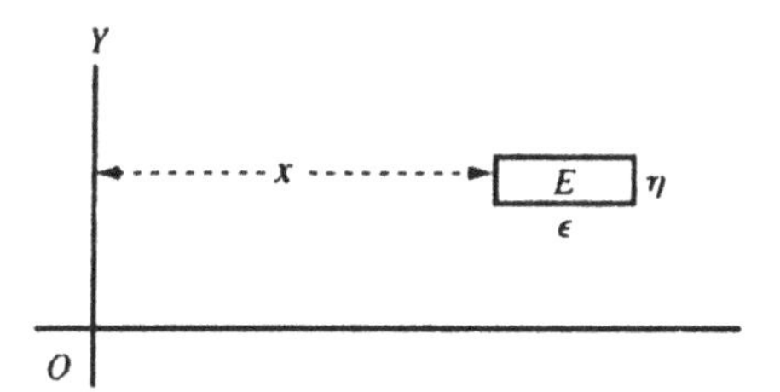

Fig. 4.1

at a distance x from the y-axis OY. The quantity $m(E) = \rho\varepsilon\eta$ is called the *mass* of E, and the quantity $\mu(E) \doteq m(E).x^2$ the *moment of inertia* of E about OY.

Now consider (Fig. 4.2) a rectangular strip S of length a and infinitesimal width η. Writing S_x for the portion of the strip of length x from OY, we shall assume that the function μ assigning S_x its moment of inertia is defined for arbitrary x and is *microadditive* in the sense that for any ε in Δ

$$\mu(S_{x+\varepsilon}) = \mu(S_x) + \mu(E_x),$$

where E_x is the rectangular microelement of S between x and $x + \varepsilon$; we also assume that $\mu(S_0) = 0$. Thus, if we define $I_0(x) := \mu(S_x)$, it follows that

$$\varepsilon I_0{}'(x) = I_0(x + \varepsilon) - I_0(x) = \mu(E_x) = \varepsilon\rho\eta x^2,$$

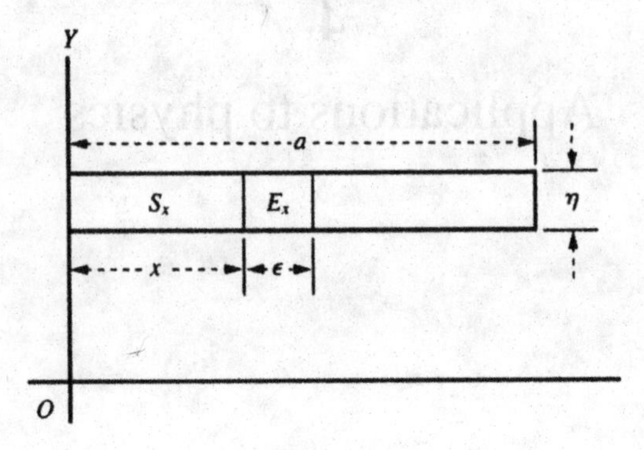

Fig. 4.2

so that, cancelling ε,

$$I_0{}'(x) = \rho\eta x^2,$$

giving (since $I_0(0) = \mu(S_0) = 0$)

$$I_0(x) = \left(\tfrac{1}{3}\right)\rho\eta x^3.$$

Accordingly, the moment of inertia $\mu(S)$ of S about OY is $\left(\frac{1}{3}\right)\rho\eta\alpha^3$ or $\left(\frac{1}{3}\right)ma^2$ where $m = \rho\eta a$ is (defined to be) the mass of S.

We use this in turn to determine the moments of inertia of a thin rectangular lamina L about various axes. Suppose L (Fig. 4.3) has sides of lengths a,b.

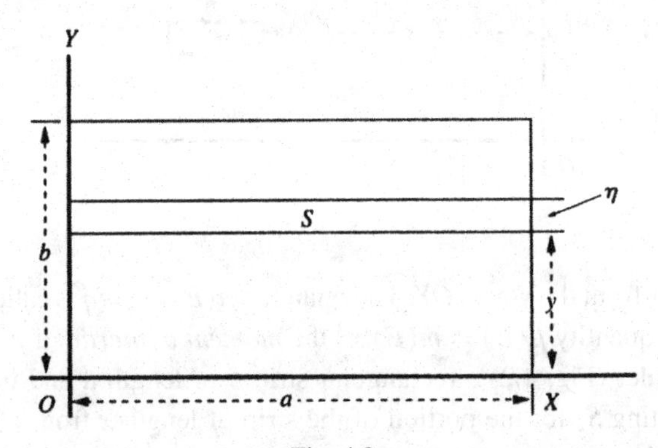

Fig. 4.3

Cut L into strips of infinitesimal width parallel to the x-axis. Let $I_1(y)$ be the moment of inertia (which we assume to be defined) about OY of the portion of L with height y. Then, as before, the assumption of 'microadditivity' of the moment of inertia function gives

$$\eta I_1{}'(y) = I_1(y + \eta) - I_1(y) = \mu(S) = \left(\tfrac{1}{3}\right)\eta\rho a^3,$$

where S is the strip bounded by the lines parallel to the x-axis at distances y, $y+\eta$ from it. Cancelling η on both sides of this equation yields $I_1{}'(y) = \left(\frac{1}{3}\right)\rho a^3$, whence $I_1(y) = \left(\frac{1}{3}\right)\rho a^3 y$, so that the moment of inertia of L about OY is

$$\left(\tfrac{1}{3}\right)\rho a^3 b = \left(\tfrac{1}{3}\right)ma^2,$$

where $m = \rho ab$ is (defined to be) the mass of L. Hence, by symmetry, the moment of inertia of L about OX is $\left(\frac{1}{3}\right)mb^2$. By translation of axes to the centre of L in accordance with the usual procedures of mechanics (see Banach, 1951), we find that the moment of inertia of L about an axis through its centre parallel to OY is

$$\left(\tfrac{1}{3}\right)ma^2 - \left(\tfrac{1}{4}\right)ma^2 = ma^2/12,$$

and the moment of inertia of L about a normal central axis is

$$\mu(L) = m(a^2 + b^2)/12.$$

This last computation is used to determine the moment of inertia of an isosceles triangular lamina. Thus consider such a lamina T (Fig. 4.4) of height

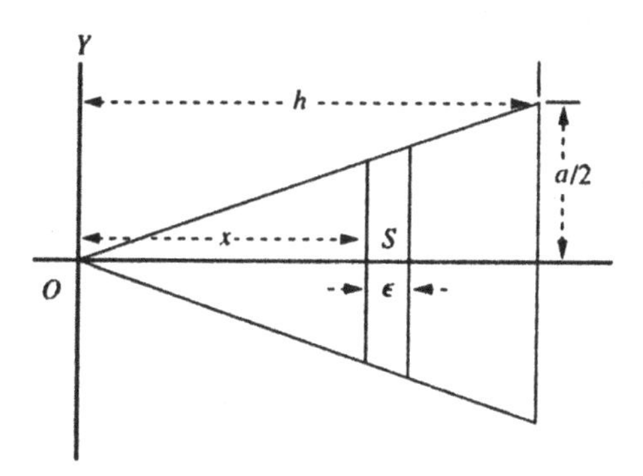

Fig. 4.4

h and base a, cut into strips of infinitesimal width parallel to its base. Let $I_2(x)$ be the moment of inertia (assumed defined) of the segment of T of height x about an axis through the origin O normal to the plane of T. For ε in Δ, let S be the strip of T between x and $x + \varepsilon$. Then, as before, 'microadditivity' of the moment of inertia function gives

$$\varepsilon I_2{}'(x) = I_2(x+\varepsilon) - I_2(x) = \text{moment of inertia of } S \text{ about } O. \qquad (4.1)$$

Now the moment of inertia $\mu(S)$ of S (Fig. 4.5) about a normal central axis is, by 'microadditivity', equal to $\mu(\nabla_1) + \mu(\nabla_2) + \mu(L)$, where ∇_1, ∇_2, L are the regions indicated in the figure. But, writing b for the slope of OP, the masses of

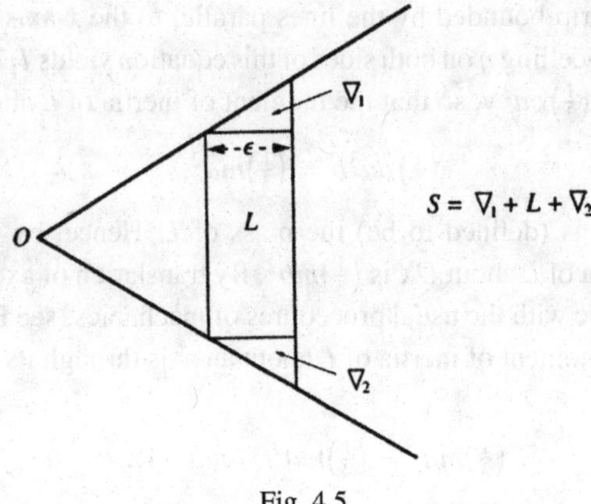

Fig. 4.5

∇_1 and ∇_2 are both equal to $\left(\frac{1}{2}\right)\rho b\varepsilon^2 = 0$. Since moment of inertia is (we shall suppose) proportional to mass, it follows that $\mu(\nabla_1) = \mu(\nabla_2) = 0$. Therefore, using the calculation for I_2,

$$\begin{aligned}\mu(S) = \mu(L) &= \text{mass } (S).(\varepsilon^2 + a^2x^2/h^2)/12 \\ &= (1/12)\,\varepsilon\rho a x/h\,.\,a^2x^2/h^2 = \varepsilon\rho a^3x^3/12h^3.\end{aligned}$$

Hence, using translation of axes and $\varepsilon^2 = 0$

$$\begin{aligned}\text{moment of inertia of } S \text{ about } O &= \varepsilon\rho a^3x^3/12h^3 + \varepsilon\rho a x(x+\varepsilon/2)^2/h \quad (4.2)\\ &= \varepsilon\rho a x^3(1 + a^2/12h^2)/h.\end{aligned}$$

Accordingly, equating (4.1) and (4.2) and cancelling ε, we obtain

$$I_2{}'(x) = \rho a x^3(1 + a^2/12h^2)/h,$$

giving

$$I_2(x) = a x^4(1 + a^2/12h^2)/4h.$$

Therefore the moment of inertia $\mu(T)$ of T about a normal axis through O is

$$\left(\tfrac{1}{4}\right)\rho a h^3(1 + a^2/12h^2) = \tfrac{1}{2}m(h^2 + a^2/12),$$

where $m = \left(\frac{1}{2}\right)\rho a h$ is (defined to be) the mass of T.

Finally we use this to determine the moment of inertia of a circular lamina about a normal central axis. Thus consider such a lamina C (Fig. 4.6) of radius r. Let $I_3(x)$ be the moment of inertia (assumed defined) about a normal axis through O of the sector OPQ, where Q has abscissa x. Let $s(x)$ be the length of the arc PQ. Now let R be the point on the circle with abscissa $x + \varepsilon$, where ε is in Δ.

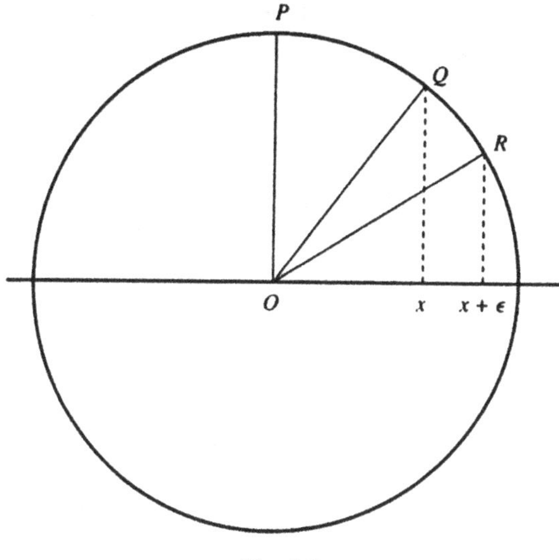

Fig. 4.6

Then, by Microstraightness, the microsegment QR of the circle is a straight line of length η, where

$$\eta = s(x + \varepsilon) - s(x) = \varepsilon s'(x).$$

Now the moment of inertia of OQR about a normal central axis through O is, as we have seen,

$$\left(\tfrac{1}{2}\right)\left(\tfrac{1}{2}\right)\eta\rho r(r^2 + \varepsilon^2/12) = \left(\tfrac{1}{4}\right)\eta\rho r^3 = \left(\tfrac{1}{4}\right)\varepsilon\rho r^3 s'(x).$$

By assuming 'microadditivity' as before, we may equate this with $I_3(x + \varepsilon) - I_3(x) = \varepsilon I_3{}'(x)$ and cancel ε to give

$$I_3{}'(x) = \rho r^3 s'(x),$$

so that

$$I_3(x) = \left(\tfrac{1}{4}\right)\rho r^3 s(x).$$

Accordingly the moment of inertia $\mu(C)$ of C about a normal central axis is $\left(\frac{1}{4}\right)\rho r^3$.circumference of C. But we have already shown that the area of C is $\left(\frac{1}{2}\right)r$.circumference of C, so that finally

$$\mu(C) = \left(\tfrac{1}{2}\right)mr^2,$$

where $m = \rho$. area of C is (defined to be) its mass.

Exercise

4.1 Use the calculation for the moment of inertia of a circular lamina to show that (i) the moment of inertia of a cylinder of mass m, height h and cross-sectional radius a about its axis is $\left(\frac{1}{2}\right)ma^2$; (ii) the moment of inertia of a sphere of mass m and radius R about a diameter is $2mr^2/5$; (iii) the moment of inertia of a right circular cone of mass m, height h and base radius a about its axis is $3ma^2/10$.

4.2 Centres of mass

The *centre of mass* of a body is the point at which the total mass of the body may be regarded as being concentrated[31]. The *centroid* of a volume, area or line is the centre of mass of a body of uniform density occupying the same space. To determine a centre of gravity or centroid one calculates the total moment of the mass about some chosen axis, as illustrated by the following example.

Suppose we want to find the centre of mass of a thin plate of uniform density and thickness in the shape of a quadrant of a circle (Fig. 4.7). Let a be the

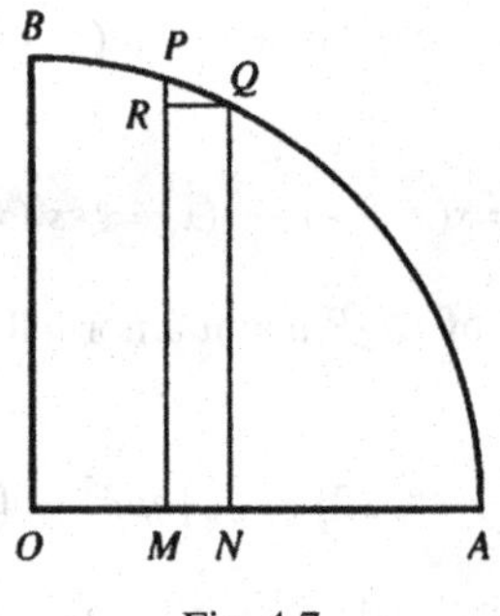

Fig. 4.7

radius of the plate and ρ the mass per unit area: the mass m of the plate is then $\left(\frac{1}{4}\right)\pi\rho a^2$. Let $\mu(x)$ be the moment about the line OA of the portion $OBPM$ of the plate, where $\underline{OM} = x$ and $\underline{MP} = y = (a^2 - x^2)^{\frac{1}{2}}$. Let Q be a point on the circle with abscissa $x + \varepsilon$, with ε in Δ. The strip $MPQN$ consists of a rectangular portion $MRQN$ together with the 'triangular defect' PQR which, being a triangle whose base and altitude are both multiples of ε, has zero area and mass. The centre of mass of $MRQN$ coincides with its geometric centre (by symmetry) and its moment about OA is $\left(\frac{1}{2}\right)y.\rho y\varepsilon = \left(\frac{1}{2}\right)\varepsilon\rho y^2$. Therefore

$$\varepsilon\mu'(x) = \mu(x+\varepsilon) - \mu(x) = \text{ moment of } MRQN = \left(\tfrac{1}{2}\right)\varepsilon\rho y^2,$$

[31] This point does not necessarily have to be contained in the body: for example, the centre of mass of a thin circular wire is, by symmetry, at its geometric centre.

so that, cancelling ε, $\mu'(x) = \left(\frac{1}{2}\right)\rho y^2 = \left(\frac{1}{2}\right)\rho(a^2 - x^2)$. Hence $\mu(x) = \left(\frac{1}{2}\right)\rho\left(a^2 x - \left(\frac{1}{3}\right)x^3\right)$, so that the total moment $\mathbf{M}$ of the quadrant about OA is

$$\mathbf{M} = \mu(a) = \left(\tfrac{1}{3}\right)\rho a^3.$$

Now, writing m for the mass of the quadrant, the y-coordinate y^* of its centre of mass is given by $my^* = \mathbf{M}$, i.e. $\left(\frac{1}{4}\right)\pi\rho a^2 y^* = \left(\frac{1}{3}\right)\rho a^3$, so that

$$y^* = 4a/3\pi.$$

By symmetry, the x-coordinate of the centre of mass of the quadrant must be the same as y^*.

Exercise

4.2 Show in the same manner as above that:

(i) the centroid of a semicircle of radius a lies at a distance $4a/3\pi$ from the centre of its diameter;

(ii) the coordinates of the centroid of a quadrant of an ellipse with axes of lengths $2a$, $2b$ are given by $x^* = 4a/3\pi$, $y^* = 4b/3\pi$;

(iii) the coordinates of the centroid of the area bounded by an arc of the parabola $y^2 = ax$, the x-axis and the line parallel to the y-axis at the point (h, k) are given by $x^* = 3h/5$, $y^* = 3k/8$;

(iv) the centroid of a right circular cone of height h is located at a distance $\left(\frac{1}{4}\right)h$ above the base.

4.3 Pappus' theorems

The concept of centroid figures in the traditional geometric facts is known as Pappus' theorems. These may be stated as follows.

I. *If an arc of a plane curve revolves about an axis in its plane which does not intersect it, the area of the surface generated by the arc is equal to the length of the arc multiplied by the length of the path of its centroid.*

II. *If a plane region revolves about an axis in its plane which does not intersect it, the volume generated by the region is equal to the area of the region multiplied by the length of the path of its centroid.*

To prove these, consider (Fig. 4.8) the closed curve CP_1P_2 – which we shall regard as a thin wire of unit density – composed of the two curves CP_1D and CP_2D, with equations $y = f_1(x)$ and $y = f_2(x)$, respectively. Let $m, m_1, m_2; M, M_1, M_2$ be the masses and the total moments about OX of the curves CP_1P_2, CP_1D and CP_2D, respectively. Then the y-coordinates y, $y_1{}^*$, $y_2{}^*$

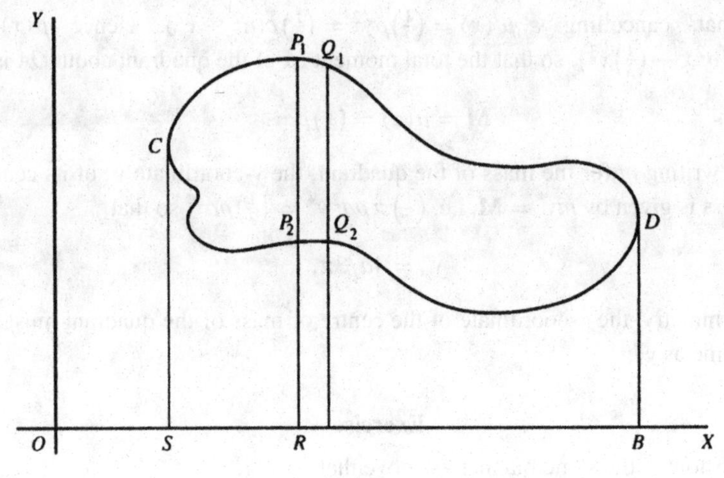

Fig. 4.8

of the centres of mass of these three curves are given by the equations

$$(m_1 + m_2)y^* = my^* = M = M_1 + M_2 \quad m_1 y_1^* = M_1 \quad m_2 y_2^* = M_2. \tag{4.3}$$

For $i = 1, 2$ let $s_i(x)$ be the length and $M_i(x)$ the moment about OX of the arc CP_i, where P_i has abscissa x. Let Q_i be a point on CP_iD with abscissa $x + \varepsilon$, with ε in Δ. Then

$$\varepsilon M_i{}'(x) = M_i(x + \varepsilon) - M_i(x) = \text{ moment of } P_i Q_i \text{ about } OX. \tag{4.4}$$

Now P_iQ_i is, by Microstraightness, a straight line of mass $s_i(x + \varepsilon) - s_i(x) = \varepsilon s_i{}'(x)$ and by symmetry its centre of mass coincides with its geometric centre, whose y-coordinate is $f_i\left(x + \left(\frac{1}{2}\right)\varepsilon\right)$. Therefore

$$\begin{aligned}\text{moment of } P_i Q_i \text{ about } OX &= \varepsilon s_i{}'(x).f_i\left(x + \left(\tfrac{1}{2}\right)\varepsilon\right) \\ &= \varepsilon s_i{}'(x)[f_i(x) + \left(\tfrac{1}{2}\right)\varepsilon f_i{}'(x)] = \varepsilon s_i{}'(x) f_i(x).\end{aligned}$$

Hence, using (4.4) and cancelling ε on both sides of the resulting equation,

$$M_i{}'(x) = f_i(x) s_i{}'(x). \tag{4.5}$$

Now, writing $S_i(x)$ for the area of the surface of revolution obtained by rotating CP_i about OX, we showed in the previous chapter that $S_i{}'(x) = 2\pi f_i(x) s_i(x)$. It follows from this and (4.5) that $2\pi M_i{}'(x) = S_i{}'(x)$, whence

$$2\pi M_i = S_i, \tag{4.6}$$

where S_i is the area of the surface of revolution generated by rotating CP_iD about OX. It now follows from (4.3) that $2\pi y_i{}^* m_i = S_i$. But m_i is numerically identical with the length s_i of CP_iD, and so we finally obtain the equation

$$2\pi y_i{}^* s_i = S_i.$$

Since $2\pi y_i{}^*$ is the distance that the centre of mass of CP_i travels in the rotation about OX, this establishes the first of Pappus' theorems for the open curves CP_1 and CP_2.

Now write s for the length of the closed curve CP_1DP_2 and S for the area of the surface it generates in rotation about OX. Then $s = s_1 + s_2$, $S = S_1 + S_2$ and so (4.3) and (4.6) give

$$\begin{aligned} 2\pi y^* &= 2\pi(s_1 + s_2)y^* = 2\pi(m_1 + m_2)y^* = 2\pi(m_1 + m_2)y^* \\ &= 2\pi(M_1 + M_2) = (S_1 + S_2) = S. \end{aligned}$$

This establishes I for the closed curve.

To obtain II, we regard the (x, y) plane, including the curve CP_1DP_2, as a flat plate of unit density. Let $M(x)$ be the moment about OX of the area CP_1P_2. Then, for ε in Δ, we have

$$\varepsilon M'(x) = M(x + \varepsilon) - M(x) = \text{moment of } P_1Q_1Q_2P_2 \text{ about } OX. \qquad (4.7)$$

Using Microstraightness in the usual way we may regard $P_1Q_1Q_2P_2$ as a rectangle whose centre of mass coincides with its geometric centre, which has y-coordinate $\left(\frac{1}{2}\right)\left[f_1\left(x + \left(\frac{1}{2}\right)\varepsilon\right) + f_2\left(x + \left(\frac{1}{2}\right)\varepsilon\right)\right]$. Since the mass of this rectangle coincides with its area $\varepsilon[f_1(x) - f_2(x)]$, it follows that

$$\begin{aligned} \text{moment of } P_1Q_1Q_2P_2 \text{ about } OX &= \left(\tfrac{1}{2}\right)[f_1\left(x + \left(\tfrac{1}{2}\right)\varepsilon\right) \\ &\quad + f_2\left(x + \left(\tfrac{1}{2}\right)\varepsilon\right)]\varepsilon[f_1(x) - f_2(x)] \\ &= \left(\tfrac{1}{2}\right)\varepsilon[f_1(x) + f_2(x)][f_1(x) - f_2(x)] \\ &= \left(\tfrac{1}{2}\right)\varepsilon[f_1(x)^2 - f_2(x)^2]. \qquad (4.8) \end{aligned}$$

Equating (4.7) and (4.8) and cancelling ε gives

$$M'(x) = \left(\tfrac{1}{2}\right)[f_1(x)^2 - f_2(x)^2]. \qquad (4.9)$$

Now let $V(x)$, $V_1(x)$, $V_2(x)$ be the volumes of revolution generated by rotating the areas CP_1P_2, SCP_1R, SCP_2R about OX. Then clearly $V(x) = V_1(x) - V_2(x)$ and it was shown in the previous chapter that $V_1{}'(x) = \pi f_1(x)^2$ and $V_2{}'(x) = \pi f_2(x)^2$. Substituting these into (4.9) gives

$$2\pi M'(x) = \pi f_1(x)^2 - \pi f(x)^2 = V_1{}'(x) - V_2{}'(x),$$

from which we deduce

$$2\pi M(x) = V_1(x) - V_2(x) = V(x).$$

It follows that, if M is the total moment about OX of the region enclosed by CP_1DP_2 and V is the volume of the surface of revolution it generates,

$$2\pi M = V. \tag{4.10}$$

Now the y-coordinate y^* of the centre of mass of the region is given by $M = my^*$, where m is its mass, which coincides numerically with its area A. Thus $M = Ay^*$. Substituting this into (4.10) gives finally

$$2\pi y^* A = V,$$

which is II.

Exercise

4.3 Use Pappus' theorems to show that
(i) the centroid of a semicircular arc of radius r lies at distance $2r/\pi$ from its centre;
(ii) the area of the surface of a torus generated by rotating a circle of radius a about an axis at distance c from its centre is $4\pi^2 ac$.

4.4 Centres of pressure

Suppose that we are given a plane area S immersed in a heavy liquid of uniform density ρ. The pressure of the liquid exerts a certain thrust on S: the *centre of pressure* in S is the point at which this thrust may be regarded as acting. To determine this point, we take moments about two lines in the plane of S, as illustrated by the following example.

Let S be a rectangle $ABCD$ (Fig. 4.9) whose plane is vertical, one side being parallel to the free surface, at which we assume the pressure is zero. Suppose $\underline{AB} = a$, $\underline{AD} = b$ and h is the depth of AB below the surface. Let $T(x)$ be the thrust on the rectangle $APQB$, where PQ is at depth x below AB. Now the pressure per unit area at depth y below the free surface is ρy. So if RS is at distance ε below PQ, with ε in Δ, the thrust on the rectangle $PQSR$ is

$$\text{average pressure per unit area over } PQSR \times \text{ area of } PQSR$$
$$= \rho(h + x + (\tfrac{1}{2})\varepsilon).a\varepsilon = \varepsilon\rho a(h + x).$$

But this thrust is also equal to $T(x+\varepsilon) - T(x) = \varepsilon T'(x)$. Equating these and cancelling ε gives $T'(x) = \rho a(h + x)$, yielding in turn

$$T(x) = \rho ahx + (\tfrac{1}{2})\rho ax^2,$$

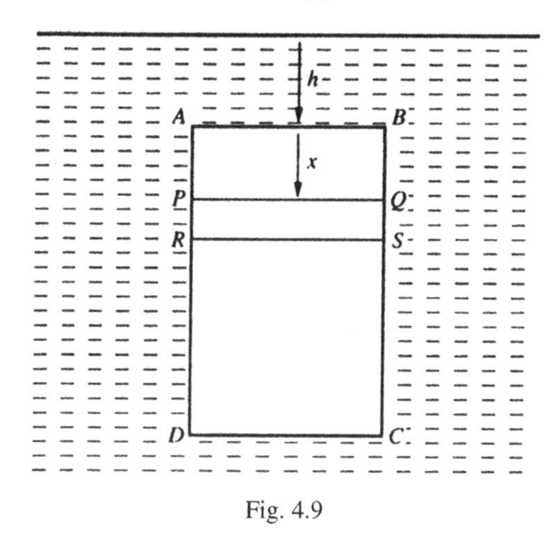

Fig. 4.9

so that the thrust on the whole rectangle *ABCD* is

$$T(b) = \rho abh + \left(\tfrac{1}{2}\right)\rho ab^2.$$

Now let $M(x)$ be the moment about *AB* of the thrust on *APQB*. The moment about *AB* of the thrust on *PQSR* is

$$\varepsilon\rho a(h + x).\left(x + \left(\tfrac{1}{2}\right)\varepsilon\right) = \varepsilon\rho ax(h + x).$$

But this moment is also equal to $M(x + \varepsilon) - M(x) = \varepsilon M'(x)$. Equating these and cancelling ε gives $M'(x) = \varepsilon\rho ax(h + x)$, so that

$$M(x) = \left(\tfrac{1}{2}\right)\rho ax^2h + \left(\tfrac{1}{3}\right)\rho ax^3.$$

Accordingly the total moment about *AB* of the thrust on *ABCD* is

$$M(b) = \left(\tfrac{1}{2}\right)\rho ab^2h + \left(\tfrac{1}{3}\right)\rho ab^3.$$

If x^* is the depth below *AB* of the centre of pressure, then this total moment must be equal to $T(b).x^*$. Thus we find

$$x^* = \left(\left(\tfrac{1}{2}\right)\rho ab^2h + \left(\tfrac{1}{3}\right)\rho ab^3\right) \div \left(\rho abh + \left(\tfrac{1}{2}\right)\rho ab^2\right) = \left(bh + \left(\tfrac{2}{3}\right)b^2\right)/(2h + b).$$

In particular, if *AB* is in the surface of the liquid, then $h = 0$ and $x^* = \left(\tfrac{2}{3}\right)b$.

Exercise

4.4 A plane lamina in the form of a parabola is lowered, with its axis vertical and its vertex downwards, into a heavy liquid. Show that, if the vertex is at depth d, the centre of pressure is at depth $4d/7$.

4.5 Stretching a spring

Let $W(x)$ be the work done in stretching a spring from its natural length a to the length $a+x$. By Hooke's law the tension $T(x)$ induced by that extension is proportional to it, so that $T(x) = Ex/a$, where E is a constant. For ε in Δ, the work done in further stretching the spring from length $a+x$ to $a+x+\varepsilon$ is given by

$$\varepsilon W'(x) = W(x+\varepsilon) - W(x) = \varepsilon T_{\text{average}},$$

where T_{average} is the average tension over the interval between $a+x$ and $a+x+\varepsilon$, that is,

$$T_{\text{average}} = \left(\tfrac{1}{2}\right)[T(x+\varepsilon) + T(x)] = T(x) + \left(\tfrac{1}{2}\right)\varepsilon T'(x).$$

Thus

$$\varepsilon W'(x) = \varepsilon\left[T(x) + \left(\tfrac{1}{2}\right)\varepsilon T'(x)\right] = \varepsilon T(x),$$

so that, cancelling ε, $W'(x) = T(x) = Ex/a$. It follows that

$$W(x) + \left(\tfrac{1}{2}\right)Ex^2/a = \left(\tfrac{1}{2}\right)xT(x).$$

4.6 Flexure of beams

On a heavy uniform beam, resting horizontally on two supports near its ends, a load is placed and as a result the beam bends slightly. We want to find an expression for the moment of the stress force acting across any given cross-section – assumed to remain plane when the beam is bent – which is originally vertical and perpendicular to the length of the beam. This will lead to an (approximate) expression for the deflection of the beam.

In accordance with the usual theory of elasticity, we assume that there is a bundle of filaments running from end to end of the beam which are neither contracted nor extended, forming a surface which we shall call the neutral surface. Let *ABCD* be a vertical section parallel to the length of the beam (see Fig. 4.10) and let *EF* be the line in which this section cuts the neutral surface. *EF* is not, by assumption, altered in length, but is curved slightly. Filaments in the section *ABDC* parallel to the original position of *EF* are bent and contracted when they lie between *EF* and *CD*, but are bent and elongated when they lie between *EF* and *AB*.

To find the stress force, consider a cross-section originally parallel to the given one and at a microdistance ε from it. After the beam is bent, both cross-sections remain plane but are inclined to each other at a microangle. Let *abc*, *def* be the lines in which they intersect the section *ABDC*, and let s be their point of intersection. Now let *gh* be a filament joining the cross-sections parallel to,

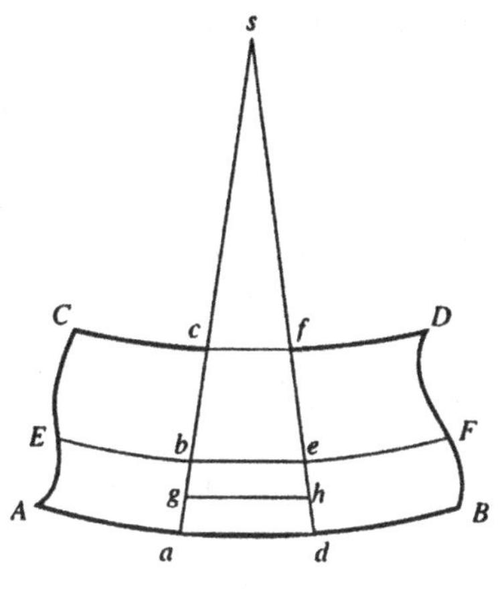

Fig. 4.10

and at distance x from, be. Before the beam is bent, gh has length ε; after the beam is bent. Microstraightness ensures that gh remains straight and parallel to be, but its length is now $(1+k)\varepsilon$, where k – its *extension* – is some constant to be determined.

Write r for the length of sb and β for the semiangle at s. Then

$$\sin\beta = \underline{gh}/2(r+x) = \varepsilon(1+k)/2(2+x).$$

Also

$$\left(\tfrac{1}{2}\right)\varepsilon = \left(\tfrac{1}{2}\right)\underline{be} = r\sin\beta = \varepsilon(1+k)/2(r+x).$$

So, cancelling ε and solving for k gives

$$k = x/r. \tag{4.11}$$

Here r is evidently the radius of curvature (see exercise 3.8) at b of the curve into which the neutral axis EF is bent.

We now take each filament (before bending) to be an extended parallelepiped of infinitesimal rectangular cross-section with sides of microlengths η,ζ. The stress force exerted on the cross-section of the filament gh at g is proportional both to its extension k and to its cross-sectional area, and so, by (4.11), equal to $Ex\,\eta\zeta/r$, where E is a constant – the *Young's modulus* of the material. Thus the moment of this stress force about the neutral axis is $Ex^2\eta\zeta/r$: in other words,

$$(E/r)\times(\text{moment of inertia of the cross-section of filament about neutral axis}).$$

In the same way as we calculated the moment of inertia of a rectangular lamina at the beginning of this chapter, it follows then that the moment of the total stress force exerted on the cross-section of the beam through abc is

$$(E/r)\times I,$$

where I is the moment of inertia of the area of the cross-section of the beam about the neutral axis.

Let us now consider a beam of uniform rectangular cross-section and length L with a load W at the middle, the weight of the beam being neglected. We assume that, when referring to the coordinates of a point on the beam, the point is always taken to lie on the unstretched neutral axis – which in this case we suppose passes through the centre of the beam – representing the curve into which the beam is bent. Take the origin of coordinates at the midpoint of the beam, the x-axis to be horizontal and the y-axis vertical, its positive direction being upwards. Let P be the point (x, y).

The bending moment M at P is the moment about the horizontal line in the cross-section through P of all the applied forces on either side of the section. The only applied force to the right of P is the reaction at the right-hand support which is $\left(\frac{1}{2}\right) W$; therefore

$$M = \left(\tfrac{1}{2}\right)W\left(\left(\tfrac{1}{2}\right)L - x\right).$$

Since the cross-section is in equilibrium, the bending moment must equal the moment of the stress exerted on it, so that

$$EI/r = \left(\tfrac{1}{2}\right)W\left(\left(\tfrac{1}{2}\right)L - x\right). \qquad (4.12)$$

By exercise 3.8, if the equation of the curve into which the beam is bent is $y = f(x)$, then the radius of curvature r at the point (x, y) is $(1 + f'^2)^{3/2}/f''$. We now make the assumption – customary in the theory of elasticity – that the bending of the beam is so slight that the square of the gradient f' is approximately equal to zero[32]. In that case we obtain (approximately) $1/r = f''$. Therefore, by (4.12),

$$EIf = \left(\tfrac{1}{2}\right)W\left(\left(\tfrac{1}{2}\right)L - x\right).$$

At the origin both f and f' are zero, so this last equation yields

$$EIf = \left(\tfrac{1}{8}\right)Wx^2\left(L - \left(\tfrac{2}{3}\right)x\right).$$

The greatest deflection f_1 occurs, say, at the beam's end, where $x = \left(\frac{1}{2}\right)L$, so that

$$f_1 = WL^3/48EI$$

[32] Here f' is not being taken as a nilsquare quantity, but merely as one whose square is 'approximately' zero. It is important to note that this is the only place in our discussion where approximations are employed.

approximately. If the section of the beam is of depth a and breadth b, then (assuming the beam to be of unit cross-sectional density), by one of our previous moment of inertia calculations,

$$I = (1/12)a^2.ab = a^3b/12.$$

So we finally obtain for the maximum deflection the approximate value

$$WL^3/4Ea^3b.$$

Therefore, to a good approximation, the deflection is proportional to the cube of the length of the beam.

4.7 The catenary, the loaded chain and the bollard-rope

A uniform flexible chain of weight w per unit length is suspended from two points A and B (Fig. 4.11). We want to determine the equation $y = f(x)$ of the curve the chain assumes.

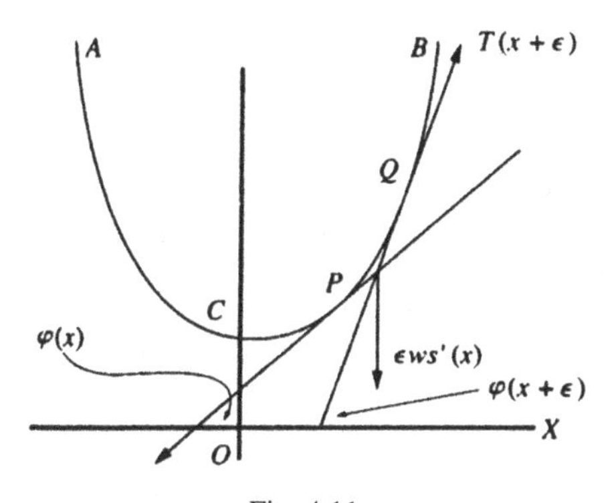

Fig. 4.11

Let $T(x)$ be the tension in the chain at a point P on it with abscissa x, let $\phi(x)$ be the angle that the tangent to the curve at P makes with the x-axis and let $s(x)$ be the length of CP. If Q is a point on the chain with abscissa $x + \varepsilon$, with ε in Δ, then the microstraight segment PQ of the chain is in equilibrium under the action of three forces, namely, its weight $w[s(x + \varepsilon) - s(x)] = \varepsilon w s'(x)$ acting vertically downwards, and the tensions $T(x)$ and $T(x + \varepsilon)$ at the ends P and Q acting in the directions of the tangents. Resolving these forces horizontally gives

$$\begin{aligned}
0 &= T(x + \varepsilon)\cos\phi(x + \varepsilon) - T\cos\phi(x) \\
&= [T(x) + \varepsilon T'(x)]\cos(\phi(x) + \varepsilon\phi'(x)) - T(x)\cos\phi(x) \\
&= [T(x) + \varepsilon T'(x)][\cos\phi(x) - \varepsilon\phi'(x)\sin\phi(x)] - T(x)\cos\phi(x) \\
&= \varepsilon T'(x)\cos\phi(x) - T(x)\phi'(x)\sin\phi(x)].
\end{aligned}$$

Hence, cancelling ε, we get

$$0 = T'(x)\cos\phi(x) - T(x)\phi'(x)\sin\phi(x) = [T(x)\cos\phi(x)]'.$$

Thus, by the Constancy Principle,

$$T(x)\cos\phi(x) = T_0, \tag{4.13}$$

where T_0 is a constant representing the horizontal component of the tension.

Next, resolving vertically, we obtain

$$\begin{aligned}\varepsilon ws'(x) &= T(x+\varepsilon)\sin\phi(x+\varepsilon) - T(x)\sin\phi(x)\\ &= [T(x)+\varepsilon T'(x)][\sin\phi(x)+\varepsilon\phi'(x)\cos\phi(x)] - T(x)\sin\phi(x)\\ &= \varepsilon[T(x)\phi'(x)\cos\phi(x) + T'(x)\sin\phi(x)].\end{aligned}$$

Hence, cancelling ε,

$$ws'(x) = T(x)\phi'(x)\cos\phi(x) + T'(x)\sin\phi(x) = [T(x)\sin\phi(x)]',$$

and so by the Constancy Principle,

$$T(x)\sin\phi(x) = ws(x). \tag{4.14}$$

From this we see that the vertical tension in the chain at any point on it is equal to its weight between that point and its lowest point.

Now recall the fundamental relation (2.7):

$$\sin\phi(x) = f'(x)\cos\phi(x).$$

This and (4.14) give

$$T(x)f'(x)\cos\phi(x) = ws(x),$$

and so, using (4.13),

$$T_0 f'(x) = ws(x).$$

Hence

$$T_0 f''(x) = ws'(x). \tag{4.15}$$

Now $s'(x) = [1+f'(x)^2]^{\frac{1}{2}}$ and substitution of this expression into (4.15) gives

$$[1+f'(x)^2]^{\frac{1}{2}} = af(x), \tag{4.16}$$

where $a = T_0/w$.

To obtain an explicit form for $f(x)$, write u for $f'(x)$ so that (4.16) becomes

$$1+u^2 = a^2u'^2. \tag{4.17}$$

We now observe that, using (4.17)

$$\begin{aligned}\left[u+(1+u^2)^{\frac{1}{2}}\right]' &= u'+uu'/(1+u^2)^{\frac{1}{2}} = \left[1+u(1+u^2)^{-\frac{1}{2}}\right]u' \\ &= \left[1+u(1+u^2)^{-\frac{1}{2}}\right](1+u^2)^{\frac{1}{2}}/a = \left[u+(1+u^2)^{\frac{1}{2}}\right]/a.\end{aligned}$$

Accordingly, writing v for $u+(1+u^2)^{\frac{1}{2}}$, we obtain the equation

$$av' = v.$$

Assuming that $u = 0$ when $x = 0$, it now follows from exercise 2.3 that $v = \exp(x/a)$, that is,

$$\exp(x/a) = u + (1+u^2)^{\frac{1}{2}}.$$

Hence

$$\exp(-x/a) = 1/\exp(a) = 1/\left[u+(1+u^2)^{\frac{1}{2}}\right] = -u+(1+u^2)^{\frac{1}{2}}.$$

Subtracting these two last equations and dividing by 2 gives

$$f'(x) = u = \left(\tfrac{1}{2}\right)[\exp(x/a) - \exp(-x/a)],$$

whence, supposing that $f(0) = a$ (that is, the lowest point of the chain lies at distance a above the x-axis),

$$y = f(x) = \left(\tfrac{1}{2}\right)a[\exp(x/a) - \exp(-x/a)].$$

This is the required equation; the corresponding curve is called the (common) *catenary.*

Exercise

4.5 Show that the tension at any point of the catenary is equal to the weight of a piece of the chain whose length is equal to the y-coordinate of the point.

Next, we determine the equation of the curve assumed by a uniformly loaded chain (or the cable of a suspension bridge). Here we assume that the ends A, B of a chain are at the same level and that the chain bears a continuously distributed load which is a constant weight w per unit run of span. Appealing to the same figure (Fig. 4.11), notation and reasoning as for the catenary, in the situation at hand we obtain the equations

$$T(x)\cos\phi(x) = T_0 \qquad T(x)\sin\phi(x) = wx.$$

Then, using the fundamental relation $\sin\phi(x) = f'(x)\cos\phi(x)$, we obtain

$$T_0 f'(x) = T(x)f'(x)\cos\phi(x) = wx.$$

It follows from this that the equation of the curve is the parabola

$$y = f(x) = wx^2/2T_0.$$

Exercise

4.6 Show that, if the span AB of a uniformly loaded chain is of length $2b$ and if the depth of the lowest point of the chain below AB is c, then the tension at B is $wb(b^2 + 4c^2)^{\frac{1}{2}}/2c$.

Finally, we determine the tension in a bollard-rope. We suppose that a rope, with free ends, presses tightly against a bollard in the form of a rough cylinder of coefficient of friction μ (Fig. 4.12). Suppose that the cord lies along the arc

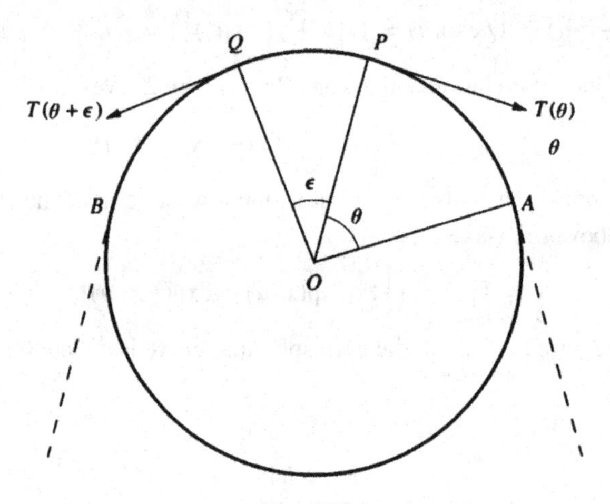

Fig. 4.12

AB in a plane normal to the axis of the cylinder. Let $T(\theta)$ be the tension at any point P on the rope between A and B, where θ is the angle between OA and OP. Let Q be a point on the rope such that OQ makes an angle ε with OP, where ε is in Δ. The tension in the rope at either end of PQ exerts a force of magnitude

$$[T(\theta + \varepsilon) + T(\theta)] \sin\left(\tfrac{1}{2}\right) \varepsilon = [2T(\theta) + \varepsilon T'(\theta)] \left(\tfrac{1}{2}\right) \varepsilon = \varepsilon T(\theta)$$

normal to PQ. Therefore, if the rope is about to slip towards A, a force of magnitude $\varepsilon\mu T(\theta)$ is exerted along PQ towards B. Since PQ is in equilibrium, the total force acting along it must be zero, and so

$$0 = T(\theta + \varepsilon) - T(\theta) + \mu\varepsilon T(\theta) = \varepsilon T'(\theta) + \mu\varepsilon T(\theta).$$

Hence, cancelling ε, and rearranging, we obtain the equation

$$T'(\theta) = -\mu T(\theta).$$

By exercise 2.3, $T(\theta)$ must then have the form

$$T(\theta) = k \exp(-\mu\theta),$$

where $k = T(0)$ is the tension at A. Thus the tension in the rope falls off exponentially.

4.8 The Kepler–Newton areal law of motion under a central force

We suppose that a particle executes plane motion under the influence of a force directed towards some fixed point O (Fig. 4.13). If P is a point on the particle's trajectory with coordinates x,y, we write r for the length of the line PO and θ for the angle that it makes with the x-axis OX. Let A be the area of the sector ORP, where R is the point of intersection of the trajectory with OX. We regard x, y, r, θ and A as functions of a time variable t: thus $x = x(t)$, $y = y(t)$, $r = r(t)$, $\theta = \theta(t)$, $A = A(t)$. Finally, let H be the acceleration towards O induced by the force: H may be a function of both r and θ.

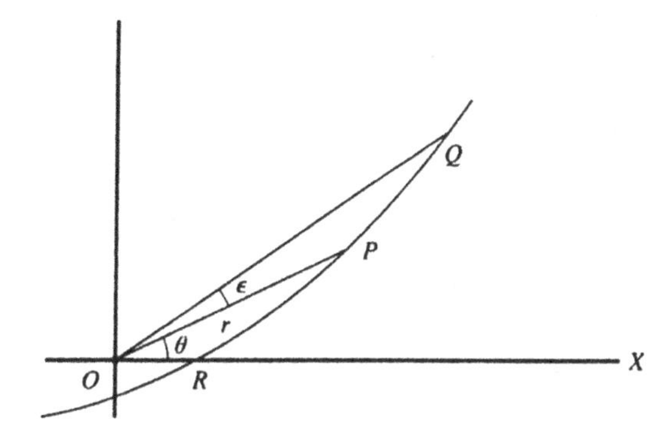

Fig. 4.13

Resolving the acceleration along and normal to OX, we have

$$x'' = H \cos\theta \qquad y'' = H \sin\theta.$$

Also $x = r\cos\theta$, $y = r\sin\theta$. Hence

$$yx'' = Hy\cos\theta = Hr\sin\theta\cos\theta \qquad xy'' = Hx\sin\theta = Hr\sin\theta\cos\theta,$$

from which we infer that

$$xy'' - yx'' = 0. \tag{4.18}$$

Now let Q be a point on the trajectory at which the time variable has value $t + \varepsilon$, with ε in Δ. Then by Microstraightness the sector OPQ is a triangle of base $r(t + \varepsilon) = r + \varepsilon r'$ and height

$$r \sin[\theta(t + \varepsilon) - \theta(t)] = r \sin \varepsilon\theta' = r\varepsilon\theta'.$$

The area of OPQ is thus

$$\left(\tfrac{1}{2}\right) \text{base} \times \text{height} = \left(\tfrac{1}{2}\right)(r + \varepsilon r').r\varepsilon\theta' = \left(\tfrac{1}{2}\right)\varepsilon r^2\theta'.$$

Therefore

$$\varepsilon A'(t) = A(t + \varepsilon) - A(t) = \text{area } OPQ = \left(\tfrac{1}{2}\right)\varepsilon r^2\theta',$$

so that, cancelling ε,

$$A'(t) = \left(\tfrac{1}{2}\right)r^2\theta'. \tag{4.19}$$

Since $x = r \cos\theta,\ y = r \sin\theta,$ we have

$$x' = r' \cos\theta - r\theta' \sin\theta \qquad y' = r' \sin\theta + r\theta' \cos\theta,$$

so that

$$yx' = rr' \sin\theta \cos\theta - r^2\theta' \sin^2\theta \qquad xy' = rr' \sin\theta \cos\theta + r^2\theta' \cos^2\theta.$$

Hence

$$xy' - yx' = r^2\theta'(\cos^2\theta + \sin^2\theta) = r^2\theta' = 2A'(t)$$

by (4.19). Then

$$2A''(t) = (xy' - yx')' = x'y' + xy'' - y'x' - yx'' = xy'' - yx'' = 0$$

by (4.18). Thus $A''(t) = 0$, so that, assuming $A(0) = 0$,

$$A(t) = kt,$$

where k is a constant.

We have therefore established the areal law of motion of a body under a central force, namely, under such a force, the radius vector joining the body to the point of origin of the force sweeps out equal areas in equal times. For planets orbiting the sun, this law was first stated by Kepler in his *Astronomica Nova* of 1609 and proved rigorously by Newton in Section II of his *Principia.*

5

Multivariable calculus and applications

5.1 Partial derivatives

Let f: $R^n \to R$ be a function $y = f(x_1, \ldots, x_{n1})$ of n variables. We define the *partial derivatives* of f as follows. For given $i = 1, \ldots, n$, fix $x_1, \ldots, x_n$ and consider the function g_i: $\Delta \to R$ defined by

$$g_i(\varepsilon) = f(x_1, \ldots, x_{i-1}, x_i + \varepsilon, x_{i+1}, \ldots, x_n).$$

Then there is, by Microaffineness, a unique b_i in R (depending on $x_1, \ldots, x_n$) such that, for all ε in Δ

$$g_i(\varepsilon) = g_i(0) + b_i.\varepsilon,$$

i.e.

$$f(x_1, \ldots, x_i + \varepsilon, \ldots, x_n) = f(x_1, \ldots, x_n) + b_i.\varepsilon.$$

The map from R^n to R which assigns b_i as defined above to each $(x_1, \ldots, x_n)$ in R^n is called the ith partial derivative of f and is written $\partial f/\partial x_i$. If f is given as a function $f(x,y,z, \ldots)$ of variables $x,y,z, \ldots$, so that x_1 is x, x_2 is $y, \ldots$, we usually write f_x for $\partial f/\partial x_1$, f_y for $\partial f/\partial x_2, \ldots$. Clearly we have

$$f(x_1, \ldots, x_i + \varepsilon, \ldots, x_n) = f(x_1, \ldots, x_n) + \varepsilon \partial f/\partial x_i(x_1, \ldots, x_n). \quad (5.1)$$

The process of forming partial derivatives may be iterated in the obvious way to obtain higher partial derivatives $\partial^2 f/\partial x_i \partial x_i$, $\partial^2 f/\partial {x_i}^2$, etc. These will also be written $f_{xy}, f_{xx}, \ldots$.

Exercises

5.1 Establish the *chain rule*: if $h = f(u(x, y, z), v(x, y, z), w(x, y, z)$, then

$$\begin{aligned}\partial h/\partial x &= (\partial f/\partial u)(\partial u/\partial x) + (\partial f/\partial v)(\partial v/\partial x)\\ &\quad + (\partial f/\partial w)(\partial w/\partial x),\end{aligned}$$

with analogous expressions for *y,z*. Generalize to functions of arbitrarily many variables.

5.2 Show that, for $f\colon R^2 \to R$, we have $f_{xy} = f_{yx}$, and generalize to functions of arbitrarily many variables. (Hint: note that, for arbitrary ε, η in Δ,

$$\begin{aligned}\eta\varepsilon f_{xy} &= f(x+\varepsilon, y+\eta) - f(x+\varepsilon, y) - [f(x, y+\eta) - f(x,y)]\\ &= f(x+\varepsilon, y+\eta) - f(x, y+\eta) - [f(x+\varepsilon, y) - f(x,y)]\\ &= \varepsilon\eta f_{yx}.)\end{aligned}$$

5.3 Suppose that $f\colon R \to R$ and $h\colon R^2 \to R$ are related by the equation

$$h(x, f(x)) = 0.$$

(We say that *h implicitly defines f.*) Show that

$$h_x + h_y f' = 0.$$

Equation (5.1) shows how the value of a multivariable function changes when one of its variables is subjected to a microdisplacement. We seek now to extend this to the case in which all of its variables are so subjected.

Let us call a pair of microquantities ε, η *proportional* if $a\varepsilon + b\eta = 0$ for some *a,b* in *R* such that $a \neq 0$ and $b \neq 0$: this means that the point (ε, η) can be joined to the origin by some definite straight line (with equation $ax + by = 0$), in other words, that (ε, η) lies in a definite direction from the origin. The region Π consisting of all proportional pairs of microquantities is a natural microneighbourhood of the origin in the plane R^2, since it is the region 'swept out' by Δ as it is allowed to rotate about the origin. Π is to be distinguished from the full Cartesian product $\Delta \times \Delta$: it is in fact easily shown that Π cannot coincide with $\Delta \times \Delta$: see below.

Evidently, if ε, η is a proportional pair, then $\varepsilon.\eta = 0$. It is this latter relation – somewhat weaker than proportionality – which we shall find most useful. So let us call a pair ε, η *mutually cancelling* if $\varepsilon\eta = 0$. We note that, by exercise 1.9(i), it is not the case that every pair of microquantities is mutually cancelling: it follows immediately from this that not every pair of microquantities is proportional.

Extending these ideas to *n*-dimensional space R^n, we shall think of an *n*-tuple of mutually cancelling microquantities (i.e. such that any pair of them is mutually cancelling) as representing a definite direction from the origin in R^n: for this reason we shall call such *n*-tuples *n*-(dimensional) *microvectors*. (Thus a 2-microvector is just a mutually cancelling pair and a 1-microvector just a microquantity.) We write $\Delta(n)$ for the collection of all *n*-microvectors; $\Delta(n)$ is called the *microspace of directions* at the origin in R^n.

Now we can extend equation (5.1).

Theorem 5.1 *Let* $f\colon R^n \to R$. *Then for any* $(x_1, \ldots, x_n)$ *in* R^n and any $(\varepsilon_1, \ldots, \varepsilon_n)$ *in* $\Delta(n)$ *we have*

$$f(x_1 + \varepsilon_1, \ldots, x_n + \varepsilon_n) = f(x_1, \ldots, x_n) + \sum_{i=1}^{n} \varepsilon_i (\partial f/\partial x_i)(x_1, \ldots, x_n).$$

Proof By induction on *n*. For $n = 1$ the assertion is a special case of (5.1). Assuming the result true for *n*, given $f\colon R^{n+1} \to R$, we fix x_{n+1} and ε_{n+1} and regard $(f(x_1, \ldots, x_n, x_{n+1} + \varepsilon_{n+1})$ as a function of $x_1, \ldots, x_n$. By inductive hypothesis,

$$\begin{aligned} f(x_1 + \varepsilon_1, \ldots, x_n + \varepsilon_n, x_{n+1} + \varepsilon_{n+1}) &= f(x_1, \ldots, x_n, x_{n+1} + \varepsilon_{n+1}) \\ &+ \sum_{i=1}^{n} \varepsilon_i (\partial f/\partial x_i)(x_i, \ldots, x_n, x_{n+1} + \varepsilon_{n+1}). \end{aligned} \tag{5.2}$$

But

$$f(x_1, \ldots, x_n, x_{n+1} + \varepsilon_{n+1}) = f(x_1, \ldots, x_{n+1}) + \varepsilon_{n+1}(\partial f/\partial x_{n+1})(x_1, \ldots, x_{n+1})$$

and

$$\begin{aligned} (\partial f/\partial x_i)(x_1, \ldots, x_n, x_{n+1} + \varepsilon_{n+1}) &= (\partial f/\partial x_i)(x_i, \ldots, x_n, x_{n+1}) \\ &+ \varepsilon_{n+1}(\partial^2 f/\partial x_{n+1}\partial x_i)(x_1, \ldots, x_n, x_{n+1}). \end{aligned}$$

Substituting these in (5.2) and recalling that the product of any pair of ε_i is zero gives

$$\begin{aligned} f(x_1 + \varepsilon_1, \ldots, x_{n+1} + \varepsilon_{n+1}) &= f(x_1, \ldots, x_{n+1}) \\ &+ \sum_{i=1}^{n+1} \varepsilon_i (\partial f/\partial x_i)(x_i, \ldots, x_{n+1}), \end{aligned}$$

completing the induction step, and the proof.

The quantity

$$\begin{aligned} \delta f = \delta f(\varepsilon_1, \ldots, \varepsilon_n) &= f(x_1 + \varepsilon_1, \ldots, x_n + \varepsilon_n) - f(x_1, \ldots, x_n) \\ &= \sum_{i=1}^{n} \varepsilon_i \partial f/\partial x_i \end{aligned}$$

is called the *microincrement* of f at $(x_1, \ldots, x_n)$ corresponding to the *n*-microvector $(\varepsilon_1, \ldots, \varepsilon_n)$. We then have

$$f(x_1 + \varepsilon_1, \ldots, x_n + \varepsilon_n) = f(x_1, \ldots, x_n) + \delta f(\varepsilon_1, \ldots, \varepsilon_n).$$

Notice that here δf represents the change in the value of f *exactly*, in contrast with the classical situation in which it represents that change only approximately.

Clearly δf may be regarded as a function – sometimes called the *differential* of f – from $R^n \times \Delta(n)$ to R given by

$$\delta f(x_1, \ldots, x_n, \varepsilon_1, \ldots, \varepsilon_n) = \sum_{i=1}^{n} \varepsilon_i(\partial f/\partial x_i)(x_1, \ldots, x_n).$$

In order to be able to apply the concept of microincrement to concrete problems we shall need to extend the Microcancellation Principle to R^n. Thus we establish the following principle.

Extended Microcancellation Principle *Given* $(a_1, \ldots, a_n)$ *in* R^n, *suppose that* $\sum_{i=1}^{n} \varepsilon_i a_i = 0$ *for any n-microvector* $(\varepsilon_1, \ldots, \varepsilon_n)$. *Then* $a_i = 0$ *for all* $i = 1, \ldots, n$.

Proof By induction on *n*. For $n = 1$ the result is (a special case of) the Microcancellation Principle. For $n = 2$, assume $\varepsilon_1 a_1 + \varepsilon_2 a_2 = 0$ for all 2-microvectors $(\varepsilon_1, \varepsilon_2)$. Noting that $(\varepsilon_1, -\varepsilon_2)$ is a 2-microvector, we can take $\varepsilon_2 = -\varepsilon_1$ to obtain $\varepsilon_1(a_1 - a_2) = 0$ for any ε_1, whence $a_1 = a_2$ by microcancellation. Therefore $2\varepsilon_1 a_1 = 0$ for any ε_1, whence $2a_1 = 0$ and so $a_1 = 0$. Thus $a_2 = a_1 = 0$.

Now, for any $n \geq 2$, assume the result true for n and suppose that $\sum_{i=1}^{n+1} \varepsilon_i a_i = 0$ for all $(n+1)$-microvectors $(\varepsilon_1, \ldots, \varepsilon_{n+1})$. Then in particular, taking $\varepsilon_{n+1} = -\varepsilon_n$,

$$0 = \sum_{i=1}^{n-1} \varepsilon_i a_i + \varepsilon(a_n - a_{n+1}).$$

It follows from the inductive hypothesis that $a_1 = a_2 = \cdots = a_{n-1} = a_n - a_{n+1} = 0$. Hence $a_n = a_{n+1}$ and we may argue as in the case $n = 2$ to conclude that $a_n = a_{n+1} = 0$. This completes the induction step and the proof.

We now turn to some applications of partial differentiation.

5.2 Stationary values of functions

If $y = f(x_1, \ldots, x_n)$ is a function from R^n to R, we shall call $(a_1, \ldots, a_n)$ a (unconstrained) *stationary point* of f and $f(a_1, \ldots, a_n)$ a *stationary value* of f if

$$f(a_1 + \varepsilon_1, \ldots, a_n + \varepsilon_n) = f(a_1, \ldots, a_n),$$

i.e. if

$$\delta f(a_1, \ldots, a_n, \varepsilon_1, \ldots, \varepsilon_n) = 0 \tag{5.3}$$

for all n-microvectors $(\varepsilon_1, \ldots, \varepsilon_n)$. In this definition we have specified n-microvectors instead of arbitrary n-tuples of microquantities because we want the value of f to be stationary under microdisplacements which lie in definite directions in R^n. Now (5.3) is the same as

$$\sum_{i=1}^{n} \varepsilon_i(\partial f/\partial x_i)(a_1, \ldots, a_n) = 0,$$

and since this is to hold for arbitrary n-microvectors $(\varepsilon_1, \ldots, \varepsilon_n)$, it follows from the Extended Microcancellation Principle that

$$(\partial f/\partial x_i)(a_1, \ldots, a_n) = 0 \qquad i = 1, \ldots, n$$

is a necessary and sufficient condition for $(a_1, \ldots, a_n)$ to be an unconstrained stationary point of f.

We next consider constrained stationary points. Thus suppose that we are given a surface S in R^n defined by the k equations (the constraint equations)

$$g_i(x_1, \ldots, x_n) = 0 \qquad i = 1, \ldots, k \tag{5.4}$$

and that we wish to determine the stationary points of a given function $f(x_1, \ldots, x_n)$ on S. We take a point $P = (x_1, \ldots, x_2)$ on S and subject it to a microdisplacement given by an n-microvector $(\varepsilon_1, \ldots, \varepsilon_n)$, so that P is displaced to $Q = (x_1 + \varepsilon_1, \ldots, x_n + \varepsilon_n)$. If Q is to remain on S, we must have

$$g_i(x_i + \varepsilon_i, \ldots, x_n + \varepsilon_n) = 0 \qquad i = 1, \ldots, k.$$

Subtracting from these the corresponding equations (5.4) gives

$$\delta g_i(\varepsilon_1, \ldots, \varepsilon_n) = 0 \qquad i = 1, \ldots, k,$$

i.e.

$$\sum_{i=1}^{n} \varepsilon_j \partial g_i/\partial x_j = 0 \qquad i = 1, \ldots, k. \tag{5.5}$$

If f is to have a stationary value at P (while constrained to remain on S), then whenever (5.4) holds we must have

$$\delta f(a_1, \ldots, a_n, \varepsilon_1, \ldots, \varepsilon_n) = 0,$$

that is

$$\sum_{j=1}^{n} \varepsilon_j \partial f/\partial x_j = 0. \tag{5.6}$$

Since (5.4) consists of k equations, we can determine k of the microquantities ε_j in terms of the remaining $n - k$, which we shall denote by $\eta_1, \ldots, \eta_{n-k}$. Substituting in (5.6) for the k thus determined ε_j in terms of $\eta_1, \ldots, \eta_{n-k}$ gives

$$\sum_{j=1}^{n-k} \eta_j h_j(x_1, \ldots, x_n) = 0$$

for some functions $h_1(x_1, \ldots, x_n), \ldots, h_{n-k}(x_1, \ldots, x_n)$. Since $(\eta_1, \ldots, \eta_{n-k})$ is an arbitrary $(n - k)$-microvector, it now follows from the Extended Microcancellation Principle that

$$h_j(x_1, \ldots, x_n) = 0 \qquad j = 1, \ldots, n - k. \tag{5.7}$$

The n equations (5.4) and (5.7) can now be solved for $x_1, \ldots, x_n$ which yield the stationary values of f subject to the given constraints.

This technique provides a nice substitute for the standard method of Lagrange multipliers employed in classical analysis. To illustrate, consider the following example.

Example Determine the volume of the largest rectangular parallelepiped inscribable in the ellipsoid

$$x^2/a^2 + y^2/b^2 + z^2/c^2 = 1 \qquad (a, b, c \neq 0).$$

Solution If x, y, z are the sides of the parallelepiped, then we seek to maximize xyz subject to the condition that the point $(x/2, y/2, z/2)$ lies on the ellipsoid, giving the constraint equation

$$x^2/a^2 + y^2/b^2 + z^2/c^2 = 4. \tag{5.8}$$

Thus the conditions for a stationary point are, for a 3-microvector $(\varepsilon, \eta, \zeta)$,

$$\varepsilon yz + \eta xz = \zeta xy = 0 \qquad \varepsilon x/a^2 + \eta y/b^2 + \zeta z/c^2 = 0.$$

Multiplying the first of these by z/c^2, the second by xy, and subtracting the results gives

$$\varepsilon(yz^2/c^2 - x^2y/a^2) + \eta(xz^2/c^2 - xy^2/b^2) = 0.$$

Applying extended microcancellation to this and cancelling y and x (since clearly both must be $\neq 0$) yields

$$z^2/c^2 - x^2/a^2 = 0 = z^2/c^2 - y^2/b^2,$$

whence

$$x = az/c, \qquad y = bz/c.$$

Substituing these in (5.8) and solving for z gives $z = 2c/\sqrt{3}$, so that the stationary (maximum) value is $xyz = 8\,abc/3\sqrt{3}$.

Exercises

5.4 Determine the dimensions of the most economical cylindrical can (with a lid) to contain n litres of water.

5.5 Find the largest volume a rectangular box can have subject to the constraint that its surface area be fixed at m square metres.

5.6 Show that the maximum value of $x^2y^2z^2$ subject to the constraint $x^2 + y^2 + z^2 = c^2$ is $c^6/27$.

5.7 A light ray travels across a boundary between two media. In the first medium its speed is v_1 and in the second it is v_2. Show that the trip is made in minimum time when *Snell's law* holds:

$$\sin\theta_1/\sin\theta_2 = v_1/v_2,$$

where θ_1, θ_2 are the angles the light ray's path make with the normal on either side of the boundary at the point of incidence.

5.3 Theory of surfaces. Spacetime metrics

Consider a surface S in R^3 defined parametrically by the equations

$$x = x(u,v) \quad y = y(u,v) \quad z = z(u,v), \tag{5.9}$$

where (u,v) ranges over some region U in R^2 (the (u,v)-plane). We think of pairs (u,v) as coordinates of points on S. If $P = (u,v)$ is a point in U, we write P_* for the corresponding point $(x(u,v), y(u,v), z(u,v))$ with 'coordinates' (u,v). We may think of (5.9) as defining a transformation of U onto S: this transformation sends each point P in U to the point P_* on S.

For fixed v_0, the function $u \to (x(u,v_0), y(u,v_0), z(u,v_0))$ defines a curve on S called the *u-curve* through v_0: clearly a point of S lies on this curve precisely when its v-coordinate is v_0. Similarly, for fixed u_0, the function $u \to (x(u_0,v), y(u_0,v), z(u_0,v))$ defines a curve on S called the *v-curve* through u_0: a point of S lies on this curve precisely when its u-coordinate is u_0. It is natural to regard the system of u- and v-curves as constituting a coordinate system (sometimes called a system of intrinsic or Gaussian coordinates: see Fig. 5.1) on S. The u- and v-curves are obtained by applying the transformation (5.9) to straight lines $v = v_0$ and $u = u_0$ in the (u,v)-plane, so that the coordinate system on S may be thought of as the result of applying the transformation (5.9) to the standard Cartesian coordinate system in the (u,v)-plane.

Suppose now we are given a curve C in U specified parametrically by the equations

$$u = u(t) \qquad v = v(t). \tag{5.10}$$

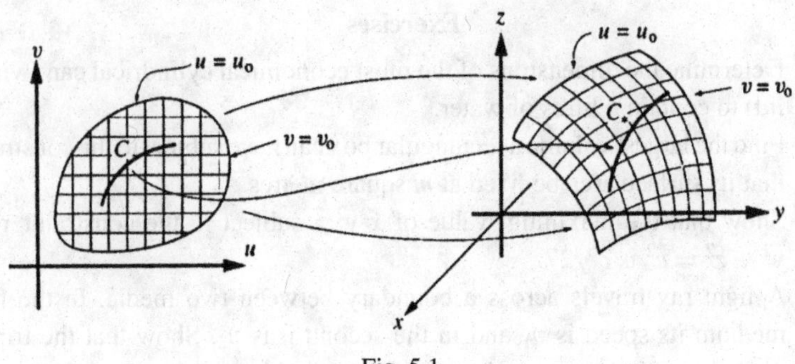

Fig. 5.1

Here t ranges over an interval in R: the point $(u(t), v(t))$ on C is said to have *curve parameter t*. Under the transformation determined by (5.9), the curve C in U is transformed into a curve C_* in S: each point P on C is mapped to the point P_* on C_*. The curve C_* is defined parametrically by the equations

$$x = x(u(t), v(t)) = x(t) \quad y = y(u(t), v(t)) = y(t)$$
$$z = z(u(t), v(t)) = z(t). \tag{5.11}$$

The point $(x(t), y(t), z(t))$ on C_* is said to have *curve parameter t*.

Let $s(t)$ be the length of C from some fixed point M to the point with curve parameter t, and let $s_*(t)$ be the length of C_* from the point M_* to the point with curve parameter t. Now consider points P, P_* on C, C_*, respectively, with common curve parameter t_0. Let Q, Q_* be the points on C, C_* with common curve parameter $t_0 + \varepsilon$, where ε is in Δ. By a result established in Chapter 3, the length of the (straight) arc PQ of C is

$$\underline{PQ} = s(t_0 + \varepsilon) - s(t_0) = \varepsilon s'(t_0) = \varepsilon \left[u'(t_0)^2 + v'(t_0)^2\right]^{\frac{1}{2}}. \tag{5.12}$$

We now determine the length of the arc P_*Q_* of C_* into which PQ is transformed. By Microstraightness, P_*Q_* is straight, and its length is $s_*(t_0 + \varepsilon) - s_*(t_0) = \varepsilon s_*'(t_0)$. Let α, β, γ be the angles that P_*Q_* make with the x, y, z-axes: then

$$\varepsilon x'(t_0) = x(t_0 + \varepsilon) - x(t_0) = \underline{P_*Q_*} \cos\alpha = \varepsilon {s_*}'(t_0)\cos\alpha,$$

so that, cancelling ε,

$$x'(t_0) = {s_*}'(t_0)\cos\alpha.$$

Similarly, $y'(t_0) = {s_*}'(t_0)\cos\beta$, $z'(t_0) = {s_*}'(t_0)\cos y$. Now $\cos\alpha, \cos\beta, \cos\gamma$, being the direction cosines of a line in R^3, are related by the equations[33]

$$\cos^2\alpha + \cos^2\beta + \cos^2\gamma = 1.$$

[33] See Hohn (1972), Chapter 3.

Therefore

$$x'(t_0)^2 + y'(t_0)^2 + z'(t_0)^2 = s_*'(t_0)^2 \left[\cos^2\alpha + \cos^2\beta + \cos^2\gamma\right] = s_*'(t_0)^2. \tag{5.13}$$

Now by the chain rule (exercise 5.1) we have, writing x' for $x'(t_0)$, x_u for $(\partial x/\partial u)(t_0)$, etc.,

$$x' = x_u u' + x_v v' \quad y' = y_u u' + y_v v' \quad z' = z_u u' + z_v v'.$$

Substituting these into (5.13) gives

$$s_*'(t_0)^2 = Eu'^2 + 2Fu'v' + Gv'^2, \tag{5.14}$$

where

$$E = x_u^2 + y_u^2 + x_u^2 \quad F = x_u x_v + y_u y_v + z_u z_v \quad G = x_v^2 + y_v^2 + z_v^2.$$

These are known as the *Gaussian fundamental quantities* of the surface *S*.

Taking the square root of both sides of (5.14) and multiplying by ε gives

$$\underline{P_*Q_*} = \varepsilon s_*'(t_0) = \varepsilon \left[Eu'^2 + 2Fu'v' + Gv'^2\right]^{\frac{1}{2}}. \tag{5.15}$$

Comparing (5.15) with (5.12), we see that the transformation (5.9) has 'stressed' (or 'shrunk') the microarc PQ of length $\varepsilon(u'^2 + v'^2)^{\frac{1}{2}}$ into the microarc P_*Q_* of length $\varepsilon\left[Eu'^2 + 2Fu'v' + Gv'^2\right]^{\frac{1}{2}}$.

Let us examine the special case in which the given curve C is a straight line passing through $(u(t_0), v(t_0) = (u_0, v_0)$ defined parametrically by the equations

$$u = u_0 + k(t - t_0) \qquad v = v_0 + \ell(t - t_0).$$

In this situation (Fig. 5.2) PQ is the diagonal of the 'coordinate microrectangle' in the (u, v)-plane bounded by the lines $u = u_0 + k\varepsilon$, $v = v_0$, $v = v_0 + \ell\varepsilon$, and P_*Q_* is the diagonal of the 'coordinate microquadrilateral' on *S* bounded by the v-curves through $u_0, u_0 + \ell\varepsilon$ and the u-curves through $v_0, v_0 + \ell\varepsilon$. The line PQ has length $\varepsilon(k^2 + \ell^2)^{\frac{1}{2}}$, while, by (5.15), since $u' = k$, $v' = \ell$, P_*Q_* has length

$$\underline{P_*Q_*} = \varepsilon Q(K, \ell)^{\frac{1}{2}}, \tag{5.16}$$

where $Q(k, \ell) = Ek^2 + 2Fk\ell + G\ell^2$ is a quadratic form in k and ℓ called the 'fundamental form' of the surface *S*. The microquantity (5.16) (whose value depends on u_0 and v_0 in addition to k and ℓ) is the magnitude of the total displacement induced by subjecting the coordinates of a location on *S* to displacements of $k\varepsilon$ and $\ell\varepsilon$ units, respectively, in the u- and v-directions. For this reason $Q(k, \ell)^{\frac{1}{2}}$ is called the *intrinsic metric* of *S*: it provides a measure of

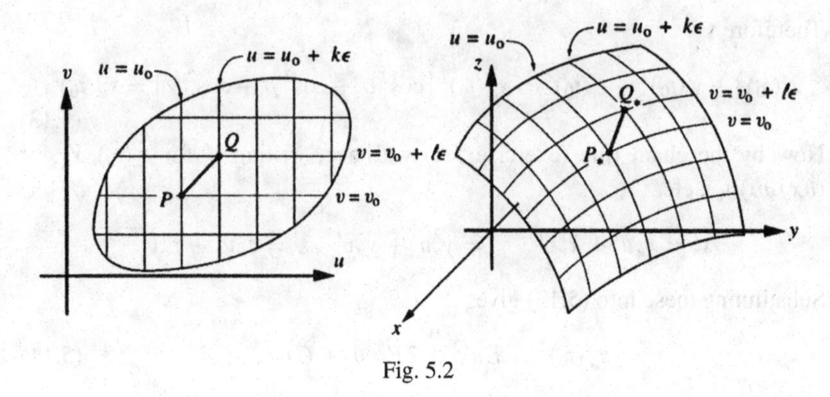

Fig. 5.2

distance – in terms of the intrinsic coordinate system – between neighbouring points on S.

Although (5.16) is a special case of (5.15) it is interesting to note that the latter can in fact be recaptured from (5.16). For, given an arbitrary curve C parametrically defined by $u = u(t)$, $v = v(t)$, by Microaffineness, we have, for any ε in Δ,

$$u(t_0 + \varepsilon) = u_0 + \varepsilon u'(t_0) \qquad v(t_0 + \varepsilon) = v_0 + \varepsilon v'(t_0).$$

Thus the arc P_*Q_* of C_* is identical with the corresponding part of the straight line

$$u(t) = u_0 + u'(t_0)(t - t_0) \qquad v(t) = v_0 + v'(t_0)(t - t_0).$$

In that case (5.16) gives

$$\underline{P_*Q_*} = \varepsilon Q(u'(t_0), v'(t_0))^{\frac{1}{2}},$$

which (aside from notational differences) is (5.15).

Finally, we return to equation (5.14). In the usual differential notation this may be written as

$$(\mathrm{d}s/\mathrm{d}t)^2 = E(\mathrm{d}u/\mathrm{d}t)^2 + 2F(\mathrm{d}u/\mathrm{d}t)(\mathrm{d}v/\mathrm{d}t) + G(\mathrm{d}v/\mathrm{d}t)^2. \qquad (5.17)$$

In classical differential geometry (see Courant, 1942, Volume II, Chapter III), it is customary to suppress the parameter t and present (5.1) in the form

$$\mathrm{d}s^2 = E\mathrm{d}u^2 + 2F\mathrm{d}u\,\mathrm{d}v + G\mathrm{d}v^2. \qquad (5.18)$$

Here ds is called the *line element* on the surface S and the expression on the right-hand side a *quadratic differential form.* Now we observe that the 'differentials' ds, du, dv in (5.18) cannot be construed as arbitrary micro-quantities in the sense of $\mathbb{S}$, since all the squared terms would reduce to zero while du dv

would, in general, not (see exercise 1.9). This being the case, what is the connection between equations (5.18) and (5.16)? We explicate it by means of an informal argument.

Think of du and dv in (5.18) as multiples ke and ℓe of some small quantity e. Then (5.18) becomes

$$\mathrm{d}s^2 = e^2\left[Ek^2 + 2Fk\ell + G\ell^2\right],$$

and so

$$\mathrm{d}s = e[Ek^2 + 2Fk\ell + G\ell^2]^{\frac{1}{2}}.$$

If we now take e to be a microquantity ε, and define $Q(k, \ell)$ as before, we obtain for the line element ds the expression

$$\varepsilon Q(k, \ell)^{\frac{1}{2}},$$

i.e. (5.16). Thus the counterpart in the smooth world $\mathbb{S}$ of the classical equation (5.18) is the equation

$$\mathrm{d}s = \varepsilon Q(k, \ell)^{\frac{1}{2}}.$$

Spacetime metrics have some arresting properties in $\mathbb{S}$. In a spacetime the metric can be written as a quadratic differential form

$$\mathrm{d}s^2 = \Sigma g_{\mu\nu}\mathrm{d}x_\mu \mathrm{d}x_\nu \qquad \mu, \nu = 1, 2, 3, 4. \tag{5.19}$$

In the classical setting (5.19) is, like (5.18), an abbreviation for an equation involving derivatives and the 'differentials' ds and dx_μ are not really quantities at all, not even microquantities. To obtain the form this equation takes in $\mathbb{S}$, we proceed as before by thinking of the dx_μ as being multiples $k_\mu e$ of some small quantity e. Then (5.19) becomes

$$\mathrm{d}s^2 = e^2\Sigma g_{\mu\nu}k_\mu k_\nu,$$

so that

$$\mathrm{d}s = e(\Sigma g_{\mu\nu}k_\mu k_\nu)^{\frac{1}{2}}.$$

Now replace e by a microquantity ε. Then we obtain the metric relation in $\mathbb{S}$:

$$\mathrm{d}s = \varepsilon(\Sigma g_{\mu\nu}k_\mu k_\nu)^{\frac{1}{2}}.$$

This tells us that the 'infinitesimal distance' ds between a point P with coordinates (x_1, x_2, x_3, x_4) and an infinitesimally near point Q with coordinates $(x_1 + k_1\varepsilon, x_2 + k_2\varepsilon, x_3 + k_3\varepsilon, x_4 + k_4\varepsilon)$ is d$s = \varepsilon(\Sigma g_{\mu\nu}k_\mu k_\nu)^{\frac{1}{2}}$. Here a curious situation arises. For when the 'infinitesimal interval' ds between P and Q

is timelike (or lightlike), the quantity $\Sigma g_{\mu\nu} k_\mu k_\nu$ is positive (or zero), so that its square root is a real number. In this case ds may be written as εd, where d is a real number. On the other hand, if ds is spacelike, then $\Sigma g_{\mu\nu} k_\mu k_\nu$ is negative, so that its square root is imaginary. In this case, then, ds assumes the form iεd, where d is a real number (and, of course $\mathrm{i} = \sqrt{-1}$). On comparing these we see that, if we take ε as the 'infinitesimal unit' for measuring infinitesimal timelike distances, then iε serves as the 'imaginary infinitesimal unit' for measuring infinitesimal spacelike distances.

For purposes of illustration, let us restrict the spacetime to two dimensions (x, t), and assume that the metric takes the simple form $\mathrm{d}s^2 = \mathrm{d}t^2 - \mathrm{d}x^2$. The infinitesimal light cone at a point P divides the infnitesimal neighbourhood at P into A timelike region T and a spacelike region S bounded by the null lines ℓ and ℓ', respectively. If we take P as origin of coordinates, a typical point Q in this neighbourhood will have coordinates $(a\varepsilon, b\varepsilon)$ with a and b real numbers: if $|b| > |a|$, Q lies in T; if $a = b$, P lies on ℓ or ℓ'; if $|a| < |b|$, p lies in S. If we write $d = |a^2 - b^2|^{\frac{1}{2}}$, then in the first case, the infinitesimal distance between P and Q is εd, in the second, it is 0, and in the third it is iεd.

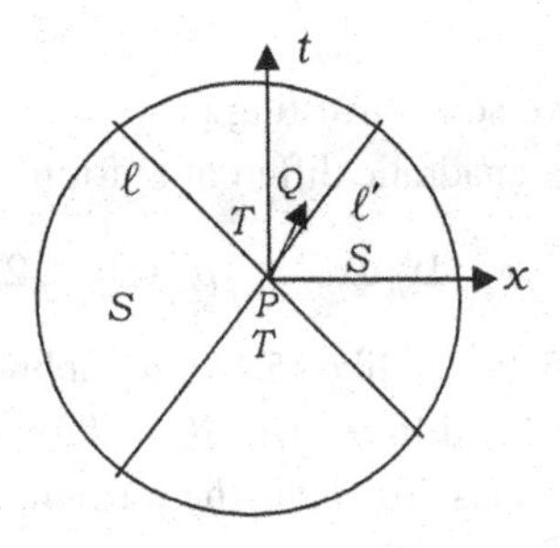

Minkowski introduced 'ict' to replace the 't' coordinate so as to make the metric of relativistic spacetime positive definite. This was purely a matter of formal convenience, and was later rejected by (general) relativists[34]. In conventional physics one never works with nilpotent quantities so it is always possible to replace formal imaginaries by their (negative) squares. But spacetime theory in $\mathbb{S}$ *forces* one to use imaginary units, since, infinitesimally, one cannot 'square oneself out of trouble'. This being the case, it would seem that, infinitesimally, the dictum *Farewell to ict*[35] needs to be replaced by

Vale 'ict', ave 'iε'!

[34] See, for example, Box 2.1. *Farewell to* 'ict', of Misner, Thorne and Wheeler (1973).
[35] See footnote 34.

To quote a well-known treaties on the theory of gravitation,

> Another danger in curved spacetime is the temptation to regard... the tangent space as lying in spacetime itself. This practice can be useful for heuristic purposes, but is incompatible with complete mathematical precision.[36]

The consistency of smooth infinitesimal analysis shows that, on the contrary, yielding to this temptation is compatible with complete mathematical precision: in $\mathbb{S}$ tangent spaces may indeed be regarded as lying in spacetime itself.

5.4 The heat equation

Suppose we are given a heated wire W (Fig. 5.3); let $T(x, t)$ be the temperature at the point P at distance x along W from some given point O on it at time t.

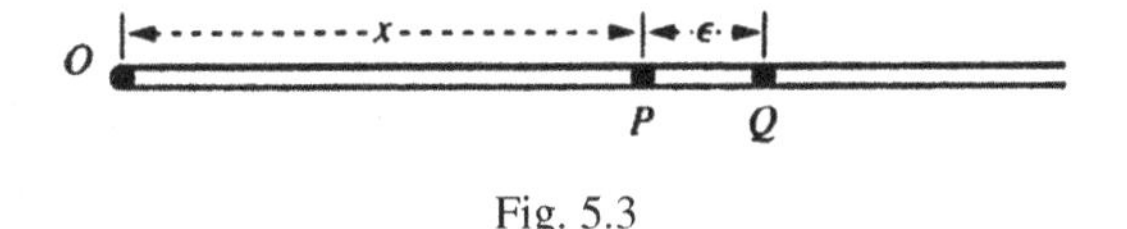

Fig. 5.3

The heat content of the segment S of W extending from x to $x + \varepsilon$, with ε in Δ, is then (by definition)

$$k\varepsilon T(x, t)$$

where k is a constant. Thus the change in heat content from time t to time $t + \eta$, with η in Δ, is

$$k\varepsilon[T(x, t + \eta) - T(x, t)] = k\varepsilon\eta T_t(x, t). \tag{5.20}$$

On the other hand, according to classical thermodynamics, the rate of flow of heat across P is proportional to the temperature gradient there, and so equal to

$$\ell T_x(x, t)$$

where ℓ is a constant. Similarly, the rate of heat flow across the point Q at distance ε from P is

$$\ell T_x(x + \varepsilon, t).$$

Thus the heat transfer across P from time t to time $t + \eta$ is

$$\ell\eta T_x(x, t),$$

and that across Q is

$$\ell\eta T_x(x + \varepsilon, t).$$

[36] *Op. cit.*, p. 205.

So the net change in heat content in S from time t to time $t+\eta$ is

$$\ell\eta\,[T_x(x+\varepsilon,t)-T_x(x,t)] = \ell\eta\varepsilon T_{xx}(x,t). \tag{5.21}$$

Equating (5.20) and (5.21) and cancelling η and ε yields the *heat equation*

$$kT_t = \ell T_{xx}.$$

5.5 The basic equations of hydrodynamics

Suppose we are given an incompressible fluid of uniform unit density flowing smoothly in space. At any point (x, y, z) in the fluid and at any time t, let

$$u = u(x,y,z,t) \quad v = v(x,y,z,t) \quad w = w(x,y,z,t)$$

be the x, y, z-components of its velocity there. Consider a volume microelement E (Fig. 5.4) at (x, y, z) which we take to be a parallelepiped with sides of length

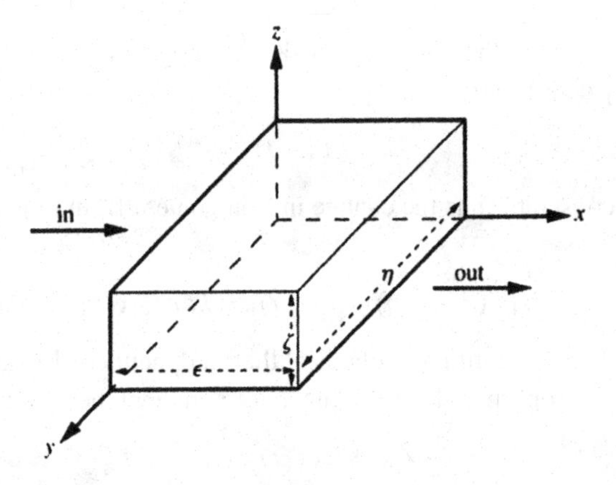

Fig. 5.4

ε, η, ζ where ε, η, ζ are arbitrary microquantities. Let us first determine the mass flow per unit time through E in the x-direction. The mass per unit time entering the left face is $u\eta\zeta$, and the mass per unit time leaving the opposite face is

$$u(x+\varepsilon, y, z)\eta\zeta = (u+\varepsilon u_x)\eta\zeta.$$

The net mass gain in the x-direction is thus $\varepsilon\eta\zeta u_x$. Similar calculations for the other directions gives the total mass gain

$$\varepsilon\eta\zeta(u_x+v_y+w_z).$$

If there is to be no creation or destruction of mass, we may equate this expression for the mass gain to zero. Since ε, η, ζ are arbitrary microquantities, they may then be cancelled, and we arrive at *Euler's equation of continuity* for an incompressible fluid:

$$u_x + v_y + w_z = 0.$$

We next need to determine the *acceleration functions* for the fluid. We define these functions

$$u^+ = u^+(x, y, z, t) \qquad v^+ = v^+(x, y, z, t) \qquad w^+ = w^+(x, y, z, t)$$

to be the rates of change of u, v, w with respect to t as we move with the fluid. That is, we want u^+, v^+, w^+ to satisfy the following conditions. Consider again the fluid element E at (x, y, z, t). For any microquantity ε, let x^ε, y^ε, z^ε be the space coordinates of E at time $t + \varepsilon$. Then we require that

$$u(x^\varepsilon, y^\varepsilon, z^\varepsilon, t + \varepsilon) = u(x, y, z, t) + \varepsilon u^+(x, y, z, t)$$
$$v(x^\varepsilon, y^\varepsilon, z^\varepsilon, t + \varepsilon) = v(x, y, z, t) + \varepsilon v^+(x, y, z, t)$$
$$w(x^\varepsilon, y^\varepsilon, z^\varepsilon, t + \varepsilon) = w(x, y, z, t) + \varepsilon w^+(x, y, z, t).$$

Now clearly, since u,v,w are the components of velocity, we have

$$x^\varepsilon = x + \varepsilon u \quad y^\varepsilon = y + \varepsilon v \quad z^\varepsilon = z + \varepsilon w.$$

Therefore

$$\varepsilon u^+ = u(x + \varepsilon u, y + \varepsilon v, z + \varepsilon w, t + \varepsilon) - u = \varepsilon(uu_x + vu_y + wu_z + u_t),$$

so that, cancelling ε,

$$u^+ = uu_x + vu_y + wu_z + u_t. \tag{5.22}$$

Similarly

$$v^+ = uv_x + vv_y + wv_z + v_t \tag{5.23}$$

$$w^+ = uw_x + vw_y + ww_z + w_t. \tag{5.24}$$

Now suppose in addition that the fluid is frictionless, and consider the forces acting on E (Fig. 5.5). Let $p = p(x, y, z, t)$ be the pressure function in the fluid. Confining attention to the x-direction, the pressure force acting on the left face of E is $\eta\zeta\, p(x, y, z, t)$, and that on the right face is

$$\eta\zeta p(x + \varepsilon, y, z, t) = \eta\zeta(p + \varepsilon p_x).$$

So the net pressure force acting on E in the positive x-direction is

$$-\varepsilon\eta\zeta p_x.$$

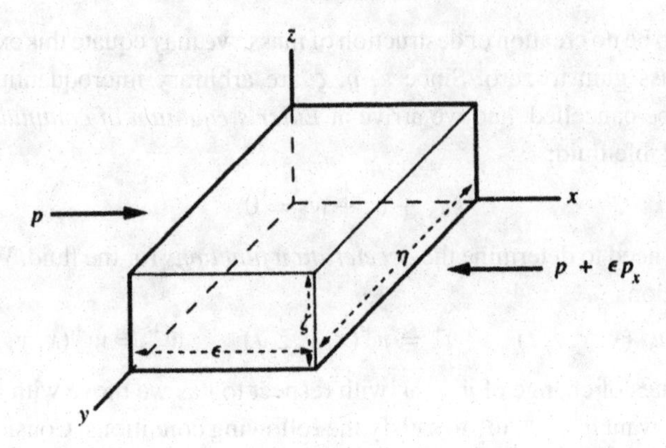

Fig. 5.5

Assuming that there are no forces acting on the fluid other than those due to pressure, then, since the force on E is the product of its mass and its acceleration, it follows that

$$-\varepsilon\eta\zeta p_x = \varepsilon\eta\zeta u^+.$$

Hence, cancelling ε, η, ζ,

$$-p_x = u^+,$$

i.e. using (5.22),

$$-p_x = uu_x + vu_y + wy_z + u_t.$$

Similarly, using (5.23) and (5.24),

$$-p_y = uv_x + vv_y + wv_z + v_t,$$

$$-p_z = uw_x + vw_y + ww_z + w_t.$$

These are *Euler's equations* for a perfect fluid.

5.6 The wave equation

Assume that the tension T and density ρ of a stretched string are both constant throughout its length (and independent of the time). Let $u(x,t)$, $\theta(x,t)$ be, respectively, the vertical displacement of the string and the angle between the string and the horizontal at position x and time t.

Consider a microelement of the string between x and $x+\varepsilon$ at time t . Its mass is $\varepsilon\rho\cos\theta(x,t)$ and its vertical acceleration $u_{tt}(x,t)$. The vertical force on the element is

$$T[\sin\theta(x+\varepsilon,t) - \sin\theta(x,t)] = \varepsilon T\theta_x(x,t)\cos\theta(x,t).$$

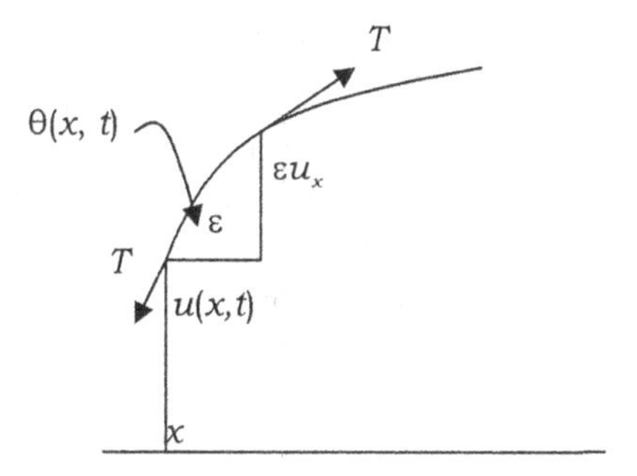

By Newton's second law, we may equate the force with mass × acceleration giving

$$\varepsilon \rho u_{tt} \cos\theta = \varepsilon T \theta_x \cos\theta.$$

Cancelling the universally quantified ε gives.

$$\rho u_{tt} \cos\theta = T \theta_x \cos\theta.$$

Since $\cos\theta \neq 0$ it may also be cancelled to give

$$\rho u_{tt} = T \theta_x. \tag{5.25}$$

Now we recall the fundamental equation governing sines and cosines (cf. (2.7)) which here takes the form

$$\sin\theta = u_x \cos\theta. \tag{5.26}$$

Applying $\dfrac{\partial}{\partial x}$ to both sides of this gives

$$\theta_x \cos\theta = -\theta_x \sin\theta + u_x \cos\theta.$$

Substituting (5.26) in this latter equation yields

$$\theta_x \cos\theta = -\theta_x u_x^2 \cos\theta + u_{xx} \cos\theta.$$

Cancelling $\cos\theta$ and rearranging gives

$$\theta_x = u_{xx}/\left(1 + u_x^2\right).$$

Substituting this in (5.25) yields the rigorous wave equation

$$u_{tt} = c^2 u_{xx}/\left(1 + u_x^2\right), \tag{5.27}$$

with $C = \sqrt{\frac{T}{P}}$.

When the amplitude of vibration is small we may assume that $u_x^2 = 0$ and in that case (5.27) becomes the familiar wave equation

$$u_{tt} = c^2 u_{xx}.$$

5.7 The Cauchy–Riemann equations for complex functions

Let $\mathbb{C}$ be the field of complex numbers (or complex plane): writing i for $\surd -1$ as usual, each complex number z is of the form $x + \mathrm{i}y$ with x, y in R. We define a *microcomplex number* to be a complex number of the form $\varepsilon + \mathrm{i}\eta$, where (ε, η) is a 2-microvector. We write Δ^* for the set of all microcomplex numbers. Clearly $z^2 = 0$ for z in Δ^*. Suppose now that we are given a function $f\colon \mathbb{C} \to \mathbb{C}$ (f is called a *complex function*). We say that f is *differentiable* at a point z in $\mathbb{C}$ if it is affine on the translate of Δ^* to z, that is, if there is a unique w in $\mathbb{C}$ such that, for all microcomplex λ,

$$f(z+\lambda) = f(z) + w\lambda. \tag{5.28}$$

We write $f'(z)$ for w. The function f is *analytic* if it is differentiable at every point of $\mathbb{C}$: in that case, the function $z \rightsquigarrow f'(z)\colon \mathbb{C} \to \mathbb{C}$ is called the *derivative* of f.

We write $u, v\colon R^2 \to R^2$ for the (functions giving the) real and imaginary parts of f; thus, for x,y, in R

$$f(x+\mathrm{i}y) = u(x,y) + \mathrm{i}v(x,y).$$

We now prove the following theorem.

Theorem 5.2 *The following conditions on a complex function f are equivalent:*

(i) *f is analytic;*

(ii) *the real and imaginary parts u, v of f satisfy the Cauchy–Riemann equations, namely,*

$$u_x = v_y \qquad v_x = -u_y.$$

Proof Suppose that (i) holds. Then for $\lambda = \varepsilon + \mathrm{i}\eta$ in Δ^* we have, writing $a + ib$ for $f'(x+\mathrm{i}y)$,

$$\begin{aligned} f((x+\mathrm{i}y)+(\varepsilon+\mathrm{i}\eta)) &= f(x+\mathrm{i}y) + (\varepsilon+\mathrm{i}\eta)(a+ib) \\ &= u(x,y) + \mathrm{i}v(x,y) + (\varepsilon a - \eta b) + \mathrm{i}(\eta a + \varepsilon b). \end{aligned} \tag{5.29}$$

But we have (independently of any assumptions),

$$\begin{aligned} f((x+\mathrm{i}y)+(\varepsilon+\mathrm{i}\eta)) &= u(x+\varepsilon, y+\eta) + \mathrm{i}v(x+\varepsilon, y+\eta) \\ &= u(x,y) + \mathrm{i}v(x,y) + \varepsilon u_x + \eta u_y + \mathrm{i}(\varepsilon v_x + \eta v). \end{aligned} \tag{5.30}$$

By equating (5.29) and (5.30), cancelling identical terms, equating real and imaginary parts and rearranging, we obtain

$$\varepsilon(u_x - a) + \eta(u_y - b) = 0 \qquad \varepsilon(v_x - b) + \eta(v_y - a) = 0.$$

Applying the Extended Microcancellation Principle to these equations gives

$$u_x = a \qquad v_x = b \qquad u_y = -b \qquad v_y = a,$$

which immediately yields (ii).

Conversely, assume (ii). To obtain (i) we have to show that, for any $z = x + \mathrm{i}y$, there is a unique $w = a + ib$ such that (5.28) holds for any microcomplex $\lambda = \varepsilon + \mathrm{i}\eta$. We first prove (independently of (ii)) the uniqueness of w. Suppose that $w_1 = a_1 + ib_1$, $w_2 = a_2 + ib_2$ both satisfy the stated condition. Then for any microcomplex $\varepsilon + i\eta$ we have

$$f(z) + (\varepsilon + \mathrm{i}\eta)(a_1 + \mathrm{i}b_1) = f(z + (\varepsilon + \mathrm{i}\eta)) = f(z) + (\varepsilon + \mathrm{i}\eta)(a_2 + \mathrm{i}b_2).$$

Cancelling $f(z)$ on both sides, multiplying out, equating real parts[37] and rearranging gives

$$\varepsilon(a_1 - a_2) + \eta(b_2 - b_1) = 0.$$

Since (ε, η) is an arbitrary 2-microvector, we may invoke extended microcancellation to infer that

$$a_1 - a_2 = b_2 - b_1 = 0,$$

whence $w_1 = w_2$. This proves the uniqueness condition.

To prove the existence of w, we note that (5.30) (which, as we observed above, holds independently of any assumptions) and (ii) give

$$\begin{aligned} f(z + \lambda) &= f((x + \mathrm{i}y) + (\varepsilon + \mathrm{i}\eta)) \\ &= [u(x, y) + \mathrm{i}v(x, y)] + \varepsilon u_x - \eta v_x + \mathrm{i}(\varepsilon v_x + \eta u_x) \\ &= f(z) + (u_x + \mathrm{i}v_x)(\varepsilon + \mathrm{i}\eta). \end{aligned}$$

So we see that (5.28) is satisfied with $w = u_x + \mathrm{i}v_x$. This gives (i), and completes the proof.

In classical complex analysis analyticity is not generally implied by satisfaction of the Cauchy–Riemann equations; one also requires that certain continuity

[37] We do not consider the imaginary part since, as is easily seen, it leads to essentially the same equation as does the real part.

conditions be satisfied. In $\mathbb{S}$, however, these extra conditions are automatically satisfied, so that the implication holds generally.

As a corollary to this theorem, we see immediately that, if f is analytic, so is its derivative f'. For if $f = u + iv$ is analytic it follows from the proof of the theorem that $f' = u_x + iv_x$, and since u and v satisfy the Cauchy–Riemann equations we have

$$u_{xx} = v_{yx} = v_{xy},$$
$$v_{xx} = -u_{yx} = -u_{xy}.$$

Thus the real and imaginary parts u_x, v_x of f' themselves satisfy the Cauchy–Riemann equations; we infer from the theorem that f' is analytic. It follows that, if a complex function is analytic, it has derivatives of arbitrarily high order.

In classical complex analysis the proof of this corollary employs complex integration: the comparatively straightforward proof we have given is made possible by the fact that, in $\mathbb{S}$, analyticity is equivalent to satisfaction of the Cauchy–Riemann equations.

Exercise

5.8 Establish, for complex functions, versions of the product and composition rules for differentiation (Chapter 2).

6

The definite integral. Higher-order infinitesimals

6.1 The definite integral

In order to be able to handle definite integrals in $\mathbb{S}$, we introduce the following principle in place of the Constancy Principle.

Integration Principle *For any f*: $[0, 1] \to R$ *there is a unique* g: $[0, 1] \to R$ *such that* $g' = f$ *and* $g(0) = 0$.

Intuitively, this principle asserts that for any f: $[0, 1] \to R$, there is a definite g: $[0, 1] \to R$ such that, for any x in $[0, 1]$, $g(x)$ is the area under the curve $y = f(x)$ from 0 to x. As usual, we write

$$\int_0^x f(t)\,\mathrm{d}t \text{ or } \int_0^x f$$

for $g(x)$, and call it a *definite integral* of f over $[0, x]$.

Exercises

6.1 Let f, g: $[0, 1] \to R$. Show that

(a) $\displaystyle\int_0^x (f + g) = \int_0^x f + \int_0^x g$,

(b) $\displaystyle\int_0^x r.f = r.\int_0^x f$,

(c) $\displaystyle\int_0^x f' = f(x) - f(0)$,

(d) $\displaystyle\int_0^x f'.g = f(x)g(x) - f(0)g(0) - \int_0^x f.g'$ (integration by parts).

(Hint: show that the two sides of each equality have the same derivative and the same value at 0.)

6.2 Let $f\colon [a, b] \times [0, 1] \to R$ and define $g(s) = \int_0^1 f(t, s)\,dt$. Show that

$$g'(s) = \int_0^1 f_s(t, s)\,dt$$

We now want to extend the definite integral to arbitrary intervals. To do this we first establish the following lemma.

Lemma 6.1 *(Hadamard) For $f\colon [a, b] \to R$ and x, y in $[a, b]$ we have*

$$f(y) - f(x) = (y - x)\int_0^1 f'(x + t(y - x))\,dt.$$

Proof For any x, y in $[a, b]$ we can define a map $h\colon [0, 1] \to [a, b]$ by $h(t) = x + t(y - x)$. Since $h' = y - x$, we have

$$f(y) - f(x) = f(h(1)) - f(h(0)) = \int_0^1 (f \circ h)'(t)\,dt = \int_0^1 (y - x)(f' \circ h)(t)\,dt$$

$$= (y - x)\int_0^1 (f' \circ h)(t)\,dt,$$

which is the required equality.

We can now prove the following theorem.

Theorem 6.2 *For any $f\colon [a, b] \to R$ there is a unique $g\colon [a, b] \to R$ such that $g' = f$ and $g(a) = 0$.*

Proof The uniqueness of g follows immediately from the observation that, if $h\colon [a, b] \to R$ and $h' = 0$, then h is constant. For if h satisfies the former condition, then by Hadamard's lemma, for x in $[a, b]$,

$$h(x) - h(a) = (x - a)\int_0^1 h'(a + t(x - a))\,dt = (x - a)\int_0^1 0\,dt = 0.$$

To establish the existence of g, define

$$g(x) = (x - a)\int_0^1 f(a + t(x - a))\,dt.$$

Clearly $g(a) = 0$ and

$$g'(x) = (x - a)'\int_0^1 f(a + t(x - a))\,dt + (x - a)\left[\int_0^1 f(a + t(x - a))\,dt\right]'.$$

Now, by exercise 6.2,

$$\left[\int_0^1 f(a+t(x-a))\,\mathrm{d}t\right]' = \int_0^1 (\partial/\partial x) f(a+t(x-a))\,\mathrm{d}t$$
$$= \int_0^1 t.f'(a+t(x-a))\,\mathrm{d}t.$$

Thus

$$g'(x) = \int_0^1 f(a+t(x-a))\,\mathrm{d}t + \int_0^1 t.f'(a-t(x-a))\,\mathrm{d}t,$$

so that, writing $h(t) = a + t(x-a)$, and using the composite rule and integration by parts,

$$\begin{aligned} g'(x) &= \int_0^1 f(h(t))\,\mathrm{d}t + \int_0^1 t.f'(h(t)).h'(t)\,\mathrm{d}t \\ &= \int_0^1 f(h(t))\,\mathrm{d}t + \int_0^1 t.(f \circ h)'(t)\,\mathrm{d}t \\ &= \int_0^1 f(h(t))\,\mathrm{d}t + [1.f(h(1)) - 0.f(h(0))] - \int_0^1 f(h(t))\,\mathrm{d}t \\ &= f(x). \end{aligned}$$

The proof is complete.

As usual, we write $\int_a^x f(t)\,\mathrm{d}t$ or $\int_a^a f$ for $g(x)$, where g is the unique function whose existence is established in this theorem.

Exercises

6.3 For f,g: $[a,b] \to R$, show that

(a) $\int_a^b f+g = \int_a^b f + \int_a^b g,$

(b) $\int_a^b r.f = r.\int_a^b f$ for r in R,

(c) $\int_a^b f' = f(b) - f(a),$

(d) $\int_a^b f'.g = f(b)g(b) - f(a)g(a) - \int_a^b f.g'.$

6.4 Let $a \leq b, c \leq d, h\colon [a,b] \to [c,d]$ with $h(a) = c, h(b) = d$, and $f\colon [c,d] \to R$. Establish the 'change of variable' formula

$$\int_c^d f(s)\,ds = \int_a^b f(h(t)).h'(t)\,dt.$$

(Hint: consider the functions

$$f_1(u) = \int_c^{h(u)} f(s)\,ds, \qquad f_2(u) = \int_a^u f(h(t)).h'(t)\,dt.$$

6.5 Show that $\int_a^x \int_b^y f(u,v)\,du\,dv = F(x,y)$ is the unique function such that $F_{xy} = f(x,y)$ and $F(x,b) = F(a,y) = 0$ for all x, y. Deduce *Fubini's theorem*:

$$\int_a^x \int_b^y f(u,v)\,du\,dv = \int_b^y \int_a^x f(u,v)\,dv\,du.$$

6.6 Let $f\colon R \to R$. Show that, if $x.f(x) = 0$ for all x in R, then $f = 0$. (Hint: for fixed but arbitrary r in R consider $h_r\colon R \to R$ defined by $h_r(x) = x.f(rx)$. Show that $h_r' = 0$.)

6.7 (a) Show that, for any $f\colon R \to R$ there is a unique $f^*\colon R \times R \to R$ such that $f(x) - f(y) = (x - y)f^*(x, y)$ for all x, y in R. (Hint: use Hadamard's lemma.)

(b) Show that $f' = f^*(x, x)$.

(c) Prove the following version of *l'Hospital's rule*. Given $f, g\colon R \to R$ with $f(0) = g(0) = 0$, suppose that $g^*(x, 0) \neq 0$ for all x in R. Then there is a unique $h\colon R \to R$ such that $f = g.h$. This function g satisfies $f'(0) = g'(0).h'(0)$. (Hint: define $h(x) = f^*(x,0)/g^*(x, 0)$. For uniqueness, use exercise 6.6.)

6.2 Higher-order infinitesimals and Taylor's theorem

If we think of those x in R such that $x^2 = 0$ (i.e. the members of Δ) as 'first-order' infinitesimals (or microquantities), then, analogously, for $k \geq 1$, the x in R for which $x^{k+1} = 0$ should be regarded as 'kth-order' infinitesimals. Let us write Δ_k for the set of all kth-order infinitesimals in this sense: observe that then $\Delta_\ell = \Delta$ and, for $k \leq \ell$, Δ_k is included in Δ_ℓ. Now the Principle of Microaffineness may be taken as asserting that any R-valued function on Δ behaves like a polynomial of degree 1. The natural extension of this to Δ_k for arbitrary k is then the assertion that any R-valued function on Δ_k behaves like a polynomial of degree k[38].

[38] Recall that in Chapter 1 we provisionally assumed that all maps on R behave locally like polynomials in order to justify the Principle of Microaffineness. Here we are finally conferring formal status on this idea.

This idea gains precise expression in the following principle whose truth in $\mathbb{S}$ we now assume.

Principle of Micropolynomiality *For any* $k \geq 1$ *and any* $g: \Delta_k \to R$, *there exist unique* $b_1, \ldots, b_k$ *in* R *such that for all* δ *in* Δ_k *we have*

$$g(\delta) = g(0) + \sum_{n=1}^{k} b_n \delta^n.$$

We may think of this as saying that Δ_k is just large enough for a function defined on it to have k derivatives, but no more. Just as the Principle of Microaffineness implied that $\Delta \neq \{0\}$, the Principle of Micropolynomiality implies that $\Delta_k \neq \Delta_{k+1}$ for any $k \geq 1$ (the simple argument establishing this is left to the reader.) The members of the Δ_k for arbitrary $k \geq 1$ are collectively known as *nilpotent infinitesimals.*

We now want to use micropolynomiality to derive a version of *Taylor's theorem*, namely, that, for any $f: R \to R$, any x in R, and any δ in Δ_k,

$$f(x + \delta) = f(x) + \sum_{n=1}^{k} \delta^n f^{(n)}(x)/n!.$$

To do this we first prove (independently of micropolynomiality) that this equality holds for all δ in Δ_k of the special form $\varepsilon_1 + \cdots + \varepsilon_k$ (by exercise 1.12, $(\varepsilon_1 + \cdots + \varepsilon_k)^{k+1} = 0$ whenever $\varepsilon_1, \ldots, \varepsilon_k$ are in Δ). We then use micropolynomiality to infer Taylor's theorem for arbitrary δ in Δ_k.

We begin with the following lemma.

Lemma 6.3 *If* $f: R \to R$ *then for any* x *in* R *and* $\varepsilon_1, \ldots, \varepsilon_k$ *in* Δ *we have*

$$f(x + \varepsilon_1 + \cdots + \varepsilon_k) = f(x) + \sum_{n=1}^{k} (\varepsilon_1 + \cdots + \varepsilon_k)^n f^{(n)}(x)/n!.$$

Proof This goes by induction on k. For $k = 1$ the assertion is just the definition of $f'(x)$. Assuming then that the assertion holds for k, we have, for $\varepsilon_1, \ldots, \varepsilon_{k+1}$ in Δ

$$\begin{aligned} & f(x + \varepsilon_1 + \cdots + \varepsilon_k + \varepsilon_{k+1}) \\ & = f(x + \varepsilon_1 + \cdots + \varepsilon_k) + \varepsilon_{k+1} f'(x + \varepsilon_1 + \cdots + \varepsilon_k) \\ & = f(x) + \sum_{n=1}^{k} (\varepsilon_1 + \cdots + \varepsilon_k)^n f^{(n)}(x)/n! \\ & \quad + \varepsilon_{k+1}[f'(x) + \sum_{n=1}^{k} (\varepsilon_1 + \cdots + \varepsilon_k)^n f^{(n+1)}(x)/n!]. \end{aligned}$$

Now the coefficient of $f^{(n)}(x)$ in the expression on the right-hand side of this last equation is, noting that, for any ε in Δ $(u+\varepsilon)^n + u^{n-1}(x+n\varepsilon)$,

$$\begin{aligned}(\varepsilon_1+\cdots+\varepsilon_k)^n/n! &+ \varepsilon_{k+1}(\varepsilon_1+\cdots+\varepsilon_k)^{n-1}/(n-1)!\\ &= (\varepsilon_1+\cdots+\varepsilon_k)^{n-1}(\varepsilon_1+\cdots+\varepsilon_k+n\varepsilon_{k+1})/n!\\ &= (\varepsilon_1+\cdots+\varepsilon_{k+1})^n/n!.\end{aligned}$$

This establishes the induction step and completes the proof.

Now, assuming micropolynomiality, we can prove the following theorem.

Theorem 6.4 (*Taylor's theorem*) *If $f: R \to R$, then for any $k \geq 1$, any x in R and any δ in Δ_k we have*

$$f(x+\delta) = f(x) + \sum_{n=1}^{k} \delta^n f^{(n)}(x)/n!.$$

Proof By micropolynomiality, for given x in R there are (unique) $b_1, \ldots, b_k$ in R such that, for all δ in Δ_k,

$$f(x+\delta) = f(x) + \sum_{n=1}^{k} b_n.\delta^n. \tag{6.1}$$

It suffices to show that $b_n = f^{(n)}(x)/n!$ for $n = 1, \ldots, k$. Taking δ in Δ_1 gives $b_1 = f'(x)$ by the definition of f'. Now suppose that $b_i = f^{(i)}(x)/i!$ for all $1 \leq i \leq n$. Then for $\varepsilon_1, \ldots, \varepsilon_{n+1}$ in Δ, $\delta = \varepsilon_1 + \cdots + \varepsilon_{n+1}$ is in Δ_{n+1} and by the lemma

$$f(x+\delta) = f(x) + \sum_{i=1}^{n} \delta^i f^{(i)}(x)/i! + \delta^{n+1} f^{(n+1)}(x)/(n+1)!.$$

But by (6.1) and the inductive hypothesis we have

$$f(x+\delta) = f(x) + \sum_{i=1}^{n} \delta^i f^{(i)}(x)/i! + b_{n+1}\,\delta^{n+1}.$$

Equating these two expressions for $f(x+\delta)$ gives

$$\delta^{n+1} b_{n+1} = \delta^{n+1} f^{(n+1)}(x)/(n+1)!. \tag{6.2}$$

But clearly $\delta^{n+1} = (\varepsilon_1+\cdots+\varepsilon_{n+1})^{n+1} = (n+1).\varepsilon_1.\varepsilon_2.\ldots.\varepsilon_{n+1}$, so that (6.2) becomes

$$\varepsilon_1.\ldots.\varepsilon_{n+1}.b_{n+1} = \varepsilon_1.\ldots.\varepsilon_{n+1}.f^{(n+1)}(x)/(n+1)!.$$

Since $\varepsilon_1, \ldots, \varepsilon_{n+1}$ are arbitrary members of Δ, they may be cancelled to give

$$b_{n+1} = f^{(n+1)}(x)/(n+1)!.$$

This completes the induction step and the proof.

Just as $\Delta = \Delta_1$ behaves as a 'universal first-order (i.e. affine) approximation' to arbitrary curves, so, analogously, Δ_k behaves as a 'universal kth-order approximation' to them. When two curves have the same kth-order approximations at a point, they are said to be in *kth-order contact* at that point. This may be made precise as follows.

Consider two curves given by $y = f(x)$ and $y = g(x)$. We may think of the images[39] $f[\Delta_k]$, $g[\Delta_k]$ as the kth-order microsegments of the two curves at 0. We say that f and g have kth-order contact at 0 if their kth-order microsegments coincide there, more precisely, if the restrictions of f and g to Δ_k are identical. By Taylor's theorem and micropolynomiality, this will be the case if and only if

$$f(0) = g(0),\ f'(0) = g'(0), \ldots, f^{(k)}(0) = g^{(k)}(0).$$

Similarly, f and g are in kth-order contact at an arbitrary point a in R if

$$f(a) = g(a),\ f'(a) = g'(a), \ldots, f^{(k)}(a) = g^{(k)}(a).$$

Exercise

6.8 Show that f and g are in second-order contact at a point a for which $f(a) = g(a)$ if and only if their tangents, curvature and osculating circles coincide there.

6.3 The three natural microneighbourhoods of zero

If we write

M_1 for the set of nilpotent infinitesimals,
M_2 for the set of x in R such that *not* $x \neq 0$,
M_3 for $[0, 0]$,

then it is readily checked that M_i is included in M_j for $i \leq j$. The M_i are called the *natural microneighbourhoods of zero*. Each determines a proximity (or 'closeness') relation on R which is reflexive, transitive and symmetric. Specifically, we define $\approx_i$ on R for $\approx_i = 1, 2, 3$ by

$$x \approx_i y \text{ if and only if } x - y \text{ is in } M_i.$$

The relations $\approx_1, \approx_2, \approx_3$ are then naturally described as the proximity relations on R determined by *algebra*, *logic*, and *order*, respectively. They are, in general, distinct, but it has been shown that it is possible for them to coincide, in other words, for M_1 to be identical with M_3 (and different from $\{0\}$).

[39] Here the image $f[X]$ of a set X under a map f is the set of all points of the form $f(x)$ with x in X.

7

Synthetic differential geometry

In this penultimate chapter we show how to formulate some of the fundamental notions of the differential geometry of manifolds in $\mathbb{S}$. In these formulations the essential objects are synthesized directly from the basic constituents of $\mathbb{S}$, avoiding the use of classical analytical methods, and thereby effecting a remarkable conceptual simplification of the foundations of differential geometry. For this reason we call differential geometry developed in $\mathbb{S}$ *synthetic*. Our account here, while purely introductory, will nevertheless enable the reader to get the gist of the approach.

Let us call the objects in $\mathbb{S}$ (e.g. R, Δ) (smooth) *spaces*. We have implicitly assumed that, for any pair of spaces S, T, we can form their product $S \times T$ in $\mathbb{S}$. We now postulate that in $\mathbb{S}$ we can form, for any spaces S,T, the space T^S of all (smooth) maps from S to T. The crucial fact about T^S is that, for any space U, there is, in $\mathbb{S}$, a bijective correspondence between maps $U \to T^S$ and maps $U \times S \to T$, which correlates each map $f: U \to T^S$ with the map $f^\wedge: U \times S \to T$ defined by

$$f^\wedge(u, s) = f(u)(s)$$

for s in S, u in U. We depict this correspondence by

$$\frac{f: U \to T^S}{f^\wedge: U \times S \to T}.$$

7.1 Tangent vectors and tangent spaces

We recall that we can think of Δ as a generic tangent vector to curves in R^2. More precisely, the space R^Δ of maps $\Delta \to R$ is, by exercise 1.8, isomorphic to R^2, which may itself be regarded as the tangent bundle of R[40]. We extend

[40] This is because each (a, b) in R^2 may be regarded as specifying a 'tangent vector' with base point a and slope b.

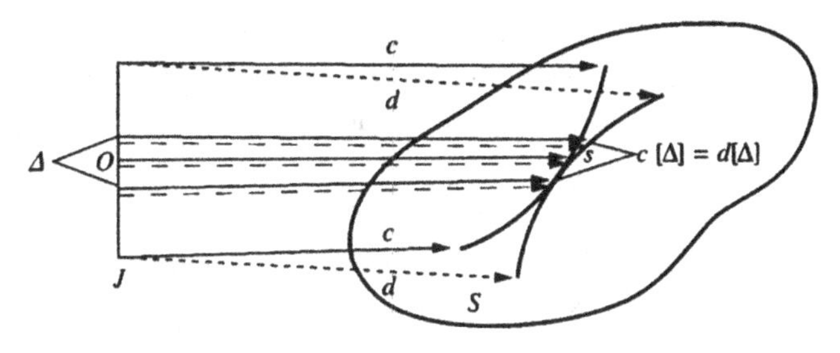

Fig. 7.1

this idea by defining, for any space S, the *tangent bundle* of S to be the space[41] S^Δ. The members of S^Δ are called *tangent vectors* to S. Associated with each tangent vector τ to S is its base point $\tau(0)$ in S: τ itself may then be thought of as a *micropath*, or as specifying a direction, in S from its base point. The *base point* map $\pi = \pi_S$: $S^\Delta \to S$ is given by $\pi(\tau) = \tau(0)$. For each point s in S we write ${S^\Delta}_s$ for $\pi^{-1}(s)$, the set of tangent vectors to S with base point s. ${S^\Delta}_s$ is called the *tangent space* (or space of micropaths) to S at s.

Here is another way of picturing tangent vectors (Fig. 7.1). For s in S, a *curve* in S with base point s is a map c: $J \to S$ with J a closed interval in R containing 0 (or R itself) such that $c(0) = s$. Call two curves c and d with common base point s *equivalent*, and write $c \approx d$, if c and d coincide on Δ[42]. Thus $c \approx d$ means that c and d are going in the 'same direction' at s, that is, are tangent at s. Clearly each tangent vector τ at s determines an equivalence class of curves with base point s, namely, the class of all such curves whose restriction to Δ coincides with τ. And conversely, any equivalence class of such curves determines a tangent vector at s, namely, the common restriction to Δ of all curves in the class. Thus tangent vectors may be identified with equivalent classes of curves in this sense[43].

Exercises

7.1 Using the fact that $R \times R$ is isomophic to R^Δ, show that the tangent bundle of R^n is isomorphic to $R^n \times R^n$.

[41] Thus the tangent bundle of any space may, in any smooth world, be regarded as a part of the space. This is in marked contrast to the classical situation in which the tangent bundle of a manifold only 'touches' the manifold.

[42] Observe that any closed interval which contains 0 also includes Δ.

[43] In classical differential geometry tangent vectors are sometimes defined as equivalence classes of curves 'pointing in the same direction' at a given point: see, Spivak (1979).

7.2 Any map $f: S \to T$ between spaces induces a map $f^\Delta: S^\Delta \to T^\Delta$ between tangent bundles given by $f^\Delta(\tau) = f \circ \tau$, for τ in S^Δ. Show that $f \circ \pi_S = \pi_T \circ f^\Delta$, and that, for each s in S, f^Δ carries ${S^\Delta}_s$ to ${T^\Delta}_{f(s)}$.

7.2 Vector fields

A *vector field* on a space S is an assignment of a tangent vector to S at each of its points, that is, a map $X: S \to S^\Delta$ such that

$$X(s)(0) = s$$

for all points s of S. This is equivalent to the condition that the composite $\pi_S \circ X$ be the identity map on S, in other words, that X is a section of π_S.

Now we know that there is a bijective correspondence of maps

$$\frac{X: S \to S^\Delta}{X^\wedge: S \times \Delta \to S}$$

with $X^\wedge(s,\varepsilon) = X(s)(\varepsilon)$ for s in S, ε in Δ, so that $X^\wedge(s,0) = s$ for all s in S. Since $X^\wedge(s, \varepsilon)$ is, intuitively, the point in S at 'distance' ε from s along the tangent vector $X(s)$, we may think of $X^\wedge$ as having the effect of moving a particle initially at s to $X^\wedge(s, \varepsilon)$. Accordingly $X^\wedge$ is called the *microflow* on S induced by X.

There is a further bijective correspondence of maps

$$\frac{X^\wedge: S \times \Delta \to S}{X^\vee: S \to S^S}$$

with $X^\vee(\varepsilon)(s) = X^\wedge(s,\varepsilon)$, In particular we have

$$X^\vee(0)(s) = X^\wedge(s,0) = s,$$

that is, $X^\vee(0)$ is the identity map on S. Thus $X^\vee$ is a micropath in the function space S^S with the identity map on S as base point. For ε in Δ the map

$$X^\vee(\varepsilon): S \to S$$

is called the *microtransformation* of S *induced by* X (and ε).

We conclude that vector fields, microflows and microtransformations are all essentially equivalent in $\mathbb{S}$. In classical differential geometry the latter two concepts cannot be rigorously defined, and figure only as suggestive metaphors.

7.3 Differentials and directional derivatives

Given $f: S \to R$, we define the map $df: S^\Delta \to R$ by

$$(df)(\tau) = (f \circ \tau)'(0)$$

for τ in S^Δ. Since, for ε in Δ,

$$f(\tau(\varepsilon)) = f(\tau(0)) + \varepsilon(f \circ \tau)'(0)$$
$$= f(\tau(0)) + \varepsilon(\mathrm{d}f)\tau,$$

it is apparent that $(\mathrm{d}f)(\tau)$ indicates the 'rate of change' of f along the micropath τ. Accordingly, df is naturally identified as the *differential* of f. Clearly

$$\delta(f \circ \tau, \varepsilon) = \varepsilon(\mathrm{d}f)(\tau),$$

where δ is the differential as defined in Chapter 5.

Exercise

7.3 For τ in ${S^\Delta}_s$ define τ^*: $R^S \to R$ by $\tau^*(f) = (\mathrm{d}f)\tau$. Show that τ^* is a *derivation* at s, i.e. satisfies $\tau^*(af + bg) = a\tau^*(f) + b\tau^*(g)$, $\tau^*(f.g) = f(s)\,\tau^*(g) + g(s)\tau^*(f)$. Thus each tangent vector gives rise to a derivation[44].

Now suppose we are given a vector field X: $S \to S^\Delta$ on S. For any f: $S \to R$ we define X^*f, the *directional derivative* of f along X, by

$$X^*f = df \circ X^\wedge.$$

Thus X^*f is a map from S to R, and from the definition of df it follows that, for ε in Δ, s in S,

$$f(X^\wedge(s,\varepsilon)) = f(s) + \varepsilon(X^*f)(s).$$

Therefore $(X^*f)(s)$ represents the rate of change of f in the direction of the microflow at s induced by the vector field X.

In particular, we may consider the vector field on R

$$\partial/\partial x\colon R \times \Delta \to R$$

given by $(x, \varepsilon) \rightsquigarrow x + \varepsilon$. In this case the directional derivative $(\partial/\partial x)\,{}^*f$ is the usual derivative f'.

Exercise

7.4 Let X be a vector field on a space S. Show that, for any f, g: $S \to R$, r in R:

(i) $X^*(r.f) = r.X^*f$,

[44] In classical differential geometry tangent vectors are sometimes defined as derivations: see Spivak (1979).

(ii) $X^*(f+g) = X^*f + X^*g$,
(iii) $X^*(f.g) = f.X^*g + g.X^*f$.

A curve c: $[a, b] \to S$ is said to be a *flow line* of the vector field X on S if

$$c(x+\varepsilon) = X^\wedge(c(x),\varepsilon)$$

for all x in $[a, b]$ and all ε in Δ. Intuitively, this means that the values of c 'go with the microflow' induced by X. If $a = b = 0$ the flow line c is called a *microflow line of* X at the point $c(0) = s$ in S; clearly in this case we have, for any ε in Δ,

$$c(\varepsilon) = X^\wedge(c(0), \varepsilon) = X^\wedge(s, \varepsilon) = X(s)(\varepsilon).$$

In other words, the restriction to Δ of any microflow line of X at a point s is $X(s)$.

Given f: $S \to R$, the curve c in S is called a *level curve* of f if the composite $f \circ c$ is constant, that is, if the value of f does not vary along c. We can now prove the following theorem relating level curves and flow lines.

Theorem 7.1 *Let X be a vector field on a space S and let f: S → R. Then the following are equivalent*:

(i) $X * f = 0$;
(ii) $f \circ X^\vee(\varepsilon) = f$ *for all* ε *in* Δ, *i.e. f is invariant under the microtransformations of S induced by X*;
(iii) *every flow line of X is a level curve of f.*

Proof The equivalence of (i) and (ii) is an immediate consequence of the definitions of X^*f and $X^\vee$. For the implication from (i) to (iii), we argue as follows. Assuming (i), we have $f(X^\wedge(s, \varepsilon)) = f(s)$ for all s in S, ε in Δ, so if c: $[a, b] \to S$ is a flow line of X, it follows that, for all x in $[a, b]$, ε in Δ,

$$f(c(x+\varepsilon)) = f(X^\wedge(x(x), \varepsilon)) = f(c(x)).$$

Therefore

$$(f \circ c)'(x) = f(c(x+\varepsilon)) - f(c(x)) = 0,$$

so that $(f \circ c)' = 0$ and $f \circ c$ is constant, i.e. c is a level curve of f.

Conversely, assuming (iii), for s in S any microflow line c for X at s is a level curve of f, and so, since the restriction of c to Δ is $X(s)$, we have, for ε in Δ,

$$f(X^\wedge(s, \varepsilon)) = f(X(s)(\varepsilon)) = f(c(\varepsilon)) = f(c(0)) = f(X(s)(0)) = f(s).$$

Hence

$$\varepsilon(X^* f)(s) = f(X^\wedge(s, \varepsilon)) - f(s) = 0,$$

and cancelling ε gives $X^* f = 0$, i.e. (i).

The proof is complete.

Exercises

7.5 A space S is said to be *microlinear* if it satisfies the following condition. For any point s in S and any n tangent vectors $\tau_1, \ldots, \tau_n$ to S at s, there is a unique map k: $\Delta(n) \to S$ such that, for all $j = 1, \ldots, n$ and all ε in Δ,

$$\tau_j(\varepsilon) = k(0, \ldots, \varepsilon, \ldots, 0) \; (\varepsilon \text{ in } j\text{th place}).$$

The *Principle of Microaffineness* for $\Delta(n)$ is the assertion that, for any map f: $\Delta(n) \to R$ there are unique $a_0, \ldots, a_n$ in R such that, for all $\varepsilon_1, \ldots, \varepsilon_n$ in Δ (n),

$$f(\varepsilon_1, \ldots, \varepsilon_n) = a_0 + \sum_{i=1}^{n} \varepsilon_i a_i.$$

Show that the Principle of Microaffineness for $\Delta(n)$ implies that R is microlinear.

7.6 Let X be a vector field on a microlinear space S. Show that, for all (ε, η) in $\Delta(2)$ and s in S,

$$X^\wedge(X^\wedge(s, \varepsilon), \eta) = X^\wedge(s, \varepsilon + \eta).$$

(Hint: use the uniqueness property in the definition of microlinearity.) Deduce that each microtransformation $X^\vee$ (ε): $S \to S$ is invertible, with $X^\vee$ $(-\varepsilon)$ as inverse. Thus any microtransformation on a microlinear space is a permutation.

Remark In a microlinear space S it is possible to define an addition operation on tangent vectors so as to make each tangent space a linear space over R, just as in classical differential geometry. To be precise, if σ and τ are two tangent vectors at the same point of S, the sum $\sigma + \tau$ is defined to be the map $\varepsilon \rightsquigarrow k(\varepsilon, \varepsilon)$, where k: $\Delta(2) \to S$ is the unique map satisfying $k(\varepsilon, 0) = \sigma(\varepsilon), k(0, \varepsilon) = \tau(\varepsilon)$ for all ε in Δ. For details see Kock (1977), Lavendhomme (1996) or Moerdijk and Reyes (1991).

8

Smooth infinitesimal analysis as an axiomatic system

At several points we have had occasion to note the fact that logic in smooth worlds differs in certain subtle respects from the classical logic with which we are familiar. These differences have not been obtrusive, and in developing the calculus and its applications in a smooth world it has not been necessary to pay particular attention to them. Nevertheless, it is a matter of logical interest to examine these differences a little more closely, and also to formulate an explicit description of the logical system which underpins reasoning in smooth worlds.

As explained in the Introduction, any smooth world $\mathbb{S}$ may be taken to be a certain type of category called a *topos*, which may be thought of as a model for mathematical concepts and operations in much the same way as the universe of set theory serves as such a model. In particular, $\mathbb{S}$ will contain an object, which we shall denote by Ω, playing the role of the set of truth values. In set theory, Ω is the set 2 consisting of two distinct individuals *true* and *false*; we assume that, in $\mathbb{S}$, Ω contains at least two such distinct individuals. Now the key property of Ω in any topos is that (just as in set theory) maps from any given object X to Ω correspond exactly to parts of X, the map with constant value *true* corresponding to X itself, and the map with constant value *false* corresponding to the empty part of X.

In set theory (or, more generally, classical logic), the usual logical operations $\wedge$ (conjunction), $\vee$ (disjunction), $\rightarrow$ (implication) and $\neg$ (negation) are defined by the familiar truth tables, so that each operation may be regarded as a map, the first three from the Cartesian product 2×2 to 2, and the last from 2 to 2. Similarly, the universal and existential quantifiers $\forall$ and $\exists$ may be thought of as being determined by the maps $\forall_X$, $\exists_X$: $2^X \rightarrow 2$, for each set X (where 2^X is the set of all maps $X \rightarrow 2$), defined by

$$\forall_X(f) = \mathit{true} \iff f(x) = \mathit{true} \text{ for all } x \in X$$

$$\exists_X(f) = \mathit{true} \iff f(x) = \mathit{true} \text{ for some } x \in X.$$

One then sees that the logical operators and quantifiers introduced in this way satisfy all the familiar rules of classical logic.

In a topos such as $\mathbb{S}$, logical operators and quantifiers can also be defined in a natural way, not by means of truth tables, but rather by exploiting the key fact that maps to Ω from objects X correspond to parts of X. For example, $\wedge$: $\Omega \times \Omega \to \Omega$ is the map corresponding to the part $\{\langle true, true\rangle\}$ of $\Omega \times \Omega$ and $\neg$: $\Omega \to \Omega$ is the map corresponding to the part $\{false\}$ of Ω. Now when we come to describe the rules satisfied by the logical operations and quantifiers introduced along these lines in a topos, we find that in general they fail to include a central rule of classical logic, the *law of excluded middle*, namely, for any proposition α,

$$\alpha \vee \neg \alpha.$$

Instead, we find that the resulting system of logical rules coincides with that of (*free*[45]) *first-order intuitionistic* or *constructive logic*. (For accounts of intuitionistic logic, see Dummett, 1977; Heyting, 1971; Kleene, 1952; or Bell and Machover, 1977.) Using standard logical notation, this has the following axioms and rules of inference:

Axioms

$$\alpha \to (\beta \to \alpha)$$
$$[\alpha \to (\beta \to \gamma)] \to [(\alpha \to \beta) \to (\alpha \to \gamma)]$$
$$\alpha \to (\beta \to \alpha \wedge \beta)$$
$$\alpha \wedge \beta \to \alpha \qquad \alpha \wedge \beta \to \beta$$
$$\alpha \to \alpha \vee \beta \qquad \beta \to \alpha \vee \beta$$
$$(\alpha \to \gamma) \to [(\beta \to \gamma) \to (\alpha \vee \beta \to \gamma)]$$
$$(\alpha \to \beta) \to [(\alpha \to \neg\beta) \to \neg\alpha]$$
$$\neg\alpha \to (\alpha \to \beta)$$
$$\alpha(t) \to \exists x \alpha(x) \quad \forall x \alpha(x) \to \alpha(t) \, (x \text{ free in, and } t \text{ free for } x \text{ in, } \alpha)$$
$$x = x \quad \alpha(x) \wedge x = y \to \alpha(y).$$

Rules of inference

$$\alpha, \alpha \to \beta / \beta \quad (\text{all variables free in } \alpha \text{ also free in } \beta)$$
$$\beta \to \alpha(x)/\beta \to \forall x \alpha(x) \quad \alpha(x) \to \beta / \exists x \alpha(x) \to \beta \, (x \text{ not free in } \beta)$$

[45] A ‘free’ logic is one whose rules are valid even when the domain of interpretation is not presumed to be nonempty.

By adding the law of excluded middle as an axiom to this system, we obtain (free) first-order *classical* logic.

It is to be noted that, not only is the law of excluded middle underivable in intuitionistic logic, so is the (equivalent) classical law of *reductio ad absurdum* or double negation

$$\neg\neg\alpha \rightarrow \alpha.$$

Nor can one derive the classical law governing the negation of the universal quantifier

$$\neg\,\forall x\alpha(x) \rightarrow \exists x\,\neg\,\alpha(x)$$

(although the reverse implication, as well as the classical law governing the negation of the existential quantifier

$$\neg\,\exists x\alpha(x) \leftrightarrow \forall x\,\neg\,\alpha(x)$$

are both intuitionistically derivable). In intuitionistic logic, existential propositions can be affirmed only when the term whose existence is asserted can be constructed or named in some definite way: it is not enough merely to assert that the assumption that all terms fail to have the property in question leads to a contradiction. In developing mathematics within a smooth world, we have not found these 'constraints' irksome because *elementary mathematics is constructive*: in practice, one always proves an existential proposition by producing an object satisfying the relevant condition.

Having specified the basic system of logic in smooth worlds, we come now to the specification of the axioms for the smooth line R therein. These are the following (which can be stated in the usual language of first-order logic).

(**R**$_1$) The structure R has specified points 0 and 1 and maps $-\colon R \rightarrow R$, $+ : R \times R \rightarrow R$ and $\cdot\colon R \times R \rightarrow R$ that make it into a nontrivial field. That is, for variables x, y, z ranging over R,

$$0 + x = x \qquad x + (-x) = 0 \qquad x + y = y + x \qquad 1.x = x \qquad x.y = y.x$$

$$(x + y) + z = x + (y + z) \quad (x.y).z = x.(y.z)$$

$$x.(y + z) = (x.y) + (x.z)$$

$$\neg\,(0 = 1)$$

$$\neg\,(x = 0) \rightarrow \exists y(x.y = 1).$$

(**R**$_2$) There is a relation $<$ on R which makes it into an ordered field in which square roots of positive elements can be extracted. That is, for variables x, y, z

ranging over R,

$$(x < y \wedge y < z) \to x < z \qquad \neg(x < x)$$

$$x < y \to x + z < y + z \qquad x < y \wedge 0 < z \to x.z < y.z$$

$$0 < 1 \qquad 0 < x \vee x < 1$$

$$0 < x \to \exists y(x = y^2)$$

$$x \neq y \to x < y \vee y < x.$$

In stating the basic axioms for smooth infinitesimal analysis we employ the usual set-theoretic notation (which is interpretable in any topos, see Bell, 1988b). Thus, for any sets X, Y, Y^X denotes the set of all maps from X to Y, $\forall x \in X\alpha(x)$ and $\exists x \in X\alpha$ (x) abbreviate 'for all x in X, $\alpha(x)$' and 'for some x in X, $\alpha(x)$', and $\{x \in X : \alpha(x)\}$ denotes the set of all x in X for which $\alpha(x)$. Define $\Delta = \{x \in R : x^2 = 0\}$. Then the basic axioms are:

($\mathbf{SIA_1}$) $$\forall f \in R^\Delta \exists! a \in R \exists! b \in R \forall \varepsilon \in \Delta . f(\varepsilon) = a + b.\varepsilon.$$

Here $\exists!$ is the usual unique existential quantifier defined by $\exists! x\alpha(x) \equiv \exists x \forall y(\alpha(y) \leftrightarrow x = y)$.

($\mathbf{SIA_2}$) $$\forall f \in R^R[\forall x \in R \forall \varepsilon \in \Delta . f(x + \varepsilon) = f(x) \to \forall x \in R \forall y \in R . f(x) = f(y)].$$

The first of these is the *Principle of Microaffineness* and the second the *Constancy Principle.*

The system comprising axioms $\mathbf{R}_1$, $\mathbf{R}_2$, $\mathbf{SIA}_1$ and $\mathbf{SIA}_2$, together with the axioms and rules of inference of free intuitionistic logic, constitutes the system **BSIA** of *basic smooth infinitesimal analysis*: we describe some of its fundamental features. To begin with, its consistency is guaranteed by the fact that topos models ('smooth worlds') have been constructed for it. Now write $\vdash_{\text{BSIA}} \alpha$ for 'α is provable in **BSIA**'. Then the proof of exercise 1.9 can be easily adapted to **BSIA** to yield

$$\vdash_{\mathbf{BSIA}} \forall f \in R^R . f \text{ is continuous},$$

where f is *continuous* is defined to mean $\forall x \in R \forall y \in R[x - y \in \Delta \to f(x) - f(y) \in \Delta]$. The proof of Theorem 1.1(ii) also adapts easily to **BSIA** to yield

$$\vdash_{\mathbf{BSIA}} \forall \varepsilon \in \Delta \neg\neg(\varepsilon = 0), \tag{8.1}$$

whence

$$\vdash_{\mathbf{BSIA}} \neg \exists \varepsilon \in \Delta \neg(\varepsilon = 0). \tag{8.2}$$

And the proof of Theorem 1.1(i) is readily adapted to **BSIA** to give

$$\vdash_{\mathbf{BSIA}} \neg\forall\varepsilon \in \Delta.\varepsilon = 0. \tag{8.3}$$

Note that of course we cannot go on to infer from (8.3) that $\vdash_{\mathbf{BSIA}} \exists\varepsilon \in \Delta \neg(\varepsilon = 0)$, since the latter, together with (8.2), would make **BSIA** inconsistent. Equations (8.2) and (8.3) together show that, while in **BSIA** it is false to suppose that 0 is the sole member of Δ, it is equally false to suppose that there exists a member of Δ which is distinct from 0. This shows how strongly the classical law governing the negation of the universal quantifier, mentioned above, can fail in intuitionistic logical systems such as **BSIA**.

From (8.1) and (8.3) it also follows that the law of excluded middle is refutable in **BSIA** in the sense that

$$\vdash_{\mathbf{BSIA}} \neg\forall x \in R[x = 0 \vee \neg(x = 0)]. \tag{8.4}$$

For, assuming $\forall x \in R[x = 0 \vee \neg(x = 0)]$, then, arguing in **BSIA**, $\forall\varepsilon \in \Delta[\varepsilon = 0 \vee \neg(\varepsilon = 0)]$. Therefore $\forall\varepsilon \in \Delta[\neg\neg(\varepsilon = 0) \rightarrow \varepsilon = 0]$. This and (8.1) now give $\forall\varepsilon \in \Delta.\varepsilon = 0$, which, with (8.3), would make **BSIA** inconsistent.

In connection with (8.4) it is interesting to note that[46], in most models of smooth infinitesimal analysis, the law of excluded middle is true in a certain restricted sense, namely, if α is any *closed* sentence, i.e. having no free variables, then $\alpha \vee \neg\alpha$ a holds. (Notice that the sentence in (8.3) is not of this form because its quantifier appears 'outside'.) Thus, in smooth infinitesimal analysis, the law of excluded middle fails 'just enough' for variables so as to ensure that all maps on R are continuous, but not so much as to affect the propositional logic of closed sentences.

The refutability of the law of excluded middle in **BSIA** leads to the refutability therein of an important principle of set theory, the *axiom of choice*. For our purposes this will be taken in the particular form

(**AC**) for any family **A** of nonempty subsets of R, there is a function $f: A \rightarrow R$ such that $f(X)$ is a member of X for any X in **A**.

We prove that **AC** is refutable in **BSIA** by showing that **AC** implies an instance of the law of excluded middle whose refutability in **BSIA** has already been established. Our argument will be informal, but easily translatable into **BSIA**.

For each $x \in R$ define

$$A_x = \{y \in R\colon y = 0 \vee (x = 0 \wedge y = 1)\},$$
$$B_x = \{y \in R\colon y = 1 \vee (x = 0 \wedge y = 0)\}.$$

[46] See McLarty (1988).

Clearly A_x and B_x are each nonempty since $0 \in A_x$ and $1 \in B_x$ for any $x \in R$. Assuming **AC**, we obtain a map f_x: $\{A_x, B_x\} \to \{0,1\}$ such that, for any $x \in R$, $f_x(A_x) \in A_x$ and $f_x(B_x) \in B_x$, in other words,

$$f_x(A_x) = 0 \vee [x = 0 \wedge f_x(A_x) = 1],$$
$$f_x(B_x) = 1 \vee [x = 0 \wedge f_x(B_x) = 0].$$

Then, using the distributive law for $\wedge$ over $\vee$ (which is valid in intuitionistic logic), we obtain

$$[f_x(A_x) = 0 \wedge f_x(B_x) = 1] \vee x = 0. \tag{8.5}$$

Now

$$x = 0 \to A_x = B_x = \{0, 1\},$$

whence

$$f_0(A_0) = f_0(B_0).$$

It follows that

$$[f_x(A_x) = 0 \wedge f_x(B_x) = 1] \to \neg(x = 0),$$

and this, together with (8.5) gives $\neg(x = 0) \vee x = 0$. Since this is the case for any $x \in R$, we obtain

$$\forall x \in R[x = 0 \vee \neg(x = 0)],$$

which, with (8.4), would make **BSIA** inconsistent.

The refutability of the axiom of choice in **BSIA**, and hence its incompatibility with the conditions of universal smoothness prevailing in smooth worlds, is not surprising in view of the axiom's well-known 'paradoxical' consequences. One of these is the famous *Banach–Tarski paradox* (see Wagon, 1985) which asserts that any solid sphere can be decomposed into finitely many pieces which can themselves be reassembled to form two solid spheres of the same size as the original. Paradoxical decompositions of this kind become possible only when smooth geometric objects are analysed into discrete sets of points which the axiom of choice then allows to be rearranged in an arbitrary (discontinuous) manner. Such procedures are not admissible in smooth worlds.

In this connection, we may also mention that the classical *intermediate value theorem*, often taken as expressing an 'intuitively obvious' property of continuous functions, is *false* in smooth worlds. This is the assertion that, for any $a, b \in R$ such that $a < b$, and any (continuous) f: $[a, b] \to R$ such that $f(a) < 0 < f(b)$, there is $x \in [a, b]$ for which $f(x) = 0$. In fact this fails even for polynomial functions in $\mathbb{S}$, as can be seen through the following informal argument. Suppose, for example, that the intermediate value theorem were true in $\mathbb{S}$ for the polynomial function $f(x) = x^3 + tx + u$. Then the value of x for which $f(x) = 0$ would have to depend smoothly on the values of t and u. In other words there would have to exist a smooth map $g: R^2 \to R$ such that

$$g(t, u)^3 + tg(t, u) + u = 0.$$

A geometric argument shows that no such smooth map can exist: for details, see Remark VII.2.14 of Moerdijk and Reyes (1991).

8.1 Natural numbers in smooth worlds

We have so far not had occasion to draw attention to the behaviour of the *natural numbers* in smooth worlds. This is because the structure of the natural number system as a whole has not played an explicit role in the basic applications we have made of smooth infinitesimal analysis. However, in certain models thereof, the system of natural numbers possesses subtle and intriguing features which, as we will see, make it possible to introduce another type of infinitesimal – the so-called *invertible* infinitesimals, which resemble those of nonstandard analysis.

To begin with, we can define the set N of *natural numbers* to be the smallest subset of R which contains 0 and is closed under the operation of adding 1. That is, writing $P(U)$ for the set of all subsets of any given set U,

$$N = \{x \in R: \forall X \in P(R)[0 \in X \wedge \forall x(x \in X \to x + 1 \in X) \to x \in X]\}.$$

Clearly then N satisfies the full induction principle, namely,

$$\forall X \in P(N)[0 \in X \wedge \forall n \in N(n \in X \to n + 1 \in X) \to X = N].$$

We can now define a set X to be *finite* if

$$\exists n \in N \exists f \in X^N. X = \{f(m): m < n\}.$$

In classical analysis the real line satisfies the *Archimedean Principle* that every real number is bounded by a natural number. Stated for R, this asserts

$$\forall x \in R \exists n \in N. x < n. \tag{8.6}$$

Moreover, in classical analysis the real line is equipped with its usual order topology which has as a base all open intervals (a, b). With this topology the real line is locally compact, i.e. every open cover of a bounded closed interval has a finite subcover. These conditions can also be stated for R in the obvious way.

Now in several models of smooth infinitesimal analysis the smooth line R resembles the classical real line in the sense of being both Archimedean and locally compact. However, models have also been constructed in which R is neither Archimedean nor locally compact. Because of this, in these models it is more natural to consider, in place of N, the set N^* of *smooth natural numbers* defined by

$$N^* = \{x \in R: x \geq 0 \wedge \sin \pi x = 0\}.$$

N^* is thus the set of points of intersection of the smooth curve $y = \sin \pi x$ with the nonnegative x-axis (Fig. 8.1). In these models R can be shown to possess the

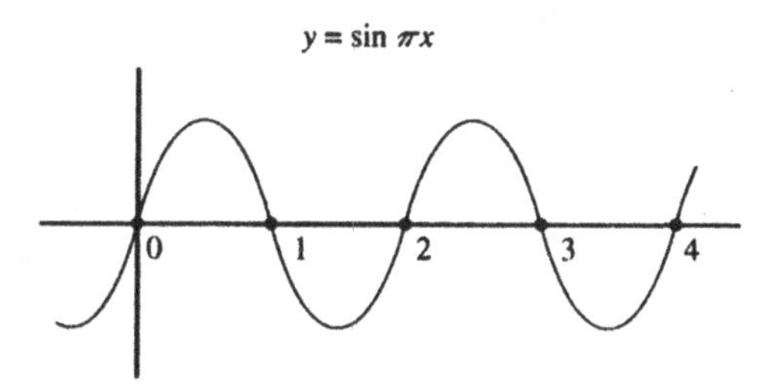

Fig. 8.1

Archimedean and local compactness properties provided that in their definitions N is replaced everywhere by N^*. (This fact enables the useful consequences of these two properties to be partially restored.) In these models, then, N and N^* do not coincide, although of course the former is a subset of the latter.

In some (but not all) models of smooth infinitesimal analysis in which R is non-Archimedean, i.e. in which (8.6) is false, it is possible to ensure that there actually exist unbounded, or infinite, elements of R in the sense of exceeding every natural number. That is, the assertion

$$\exists x \in R \forall n \in N . n < x \tag{8.7}$$

becomes true in these models. Now from $\forall n \in N . n < x$ it follows that

$$\forall n \in N . -1/(n+1) < 1/x < 1/(n+1),$$

and $\neg(1/x = 0)$ since clearly $x > 0$. Thus (8.7) implies that

$$\exists x \in R[\neg (x = 0) \wedge \forall n \in N(-1/(n+1) < x < 1/(n+1))]. \tag{8.8}$$

Members of the set

$$I = \{x \in R: \neg (x = 0) \wedge \forall n \in N(-1/(n+1) < x < 1/(n+1)\}$$

are called *invertible infinitesimals*: they are the multiplicative inverses of 'infinite' elements of R.

Exercises

8.1 Define $\Gamma = \{x \in R: \forall n \in N(-1/(n+1) < x < 1/(n+1))\}$: the members of Γ are called *infinitely small*. Show that

$$x \leq 0 \wedge 0 \leq x \rightarrow x \in \Gamma,$$

and deduce that Γ includes all the microneighbourhoods M_1, M_2, M_3 defined in Chapter 6. That is, all infinitesimals of smooth infinitesimal analysis are infinitely small in this sense.

8.2 Define $\Gamma^* = \{x \in R: \forall n \in N^*(-1/(n+1) < x < 1/(n+1)\}$. Assuming that R is *smoothly Archimedean*, i.e. $\forall x \in R \exists n \in N^*(x < n)$, show that $\Gamma^* = M_3$.

Models of smooth infinitesimal analysis in which (8.8) holds are said to contain invertible infinitesimals[47] (in addition to the nilpotent ones). The presence of invertible infinitesimals is a central feature of the theory of infinitesimals known as *nonstandard analysis*. We conclude with a thumbnail sketch of that theory, contrasting it with smooth infinitesimal analysis.

8.2 Nonstandard analysis

In nonstandard analysis one starts with the classical real line $\mathbb{R}$ and considers a 'universe' over it: here by a universe is meant a set U containing $\mathbb{R}$ which is closed under the usual set-theoretic operations of union, power set, Cartesian products and subsets. Let $\mathbf{U}$ be the structure $(U, \in)$, where $\in$ is the usual membership relation on U: associated with this is a first-order language $\mathbf{L}(\mathbf{U})$ containing a binary predicate symbol, also written $\in$, and a name for each element of U. The *restricted sentences* of this language are those in which each quantifier is of the form $\forall x \in \mathbf{u}$ or $\exists x \in \mathbf{u}$, where $\mathbf{u}$ is the name for the element u of U. Using the well-known compactness theorem for first-order logic, a new set $^*U$, called an *enlargement* of U, and a map $u \rightarrow {}^*u$: $U \rightarrow {}^*U$ are constructed so as to possess the following properties:

i. The map $*$ is a restricted elementary embedding of $\mathbf{U}$ into the structure $^*\mathbf{U} = (^*U, \in)$, that is, for any restricted sentence α of $\mathbf{L}(\mathbf{U})$, α holds in $\mathbf{U}$ if

[47] The principal use to which invertible infinitesimals have been put in smooth infinitesimal analysis is in the theory of distributions and the construction of the Dirac δ-function, topics which are too advanced to be treated in an elementary book of this kind. See Chapter VII of Moerdijk and Reyes (1991).

and only if it holds in $^*\mathbf{U}$ when each name $\mathbf{u}$ occurring in α is interpreted as the element $^*u$ of $^*U$.

ii. The set $^*\mathbb{R}$ properly includes the set $\{^*r: r \in \mathbb{R}\}$.

It follows from (i) that $^*\mathbb{R}$ has identical set-theoretical properties (i.e. expressible in terms $\in$) as does $\mathbb{R}$ and is therefore a model of the classical theory of real numbers. However, if we identify each $r \in \mathbb{R}$ with its image $^*r$ in $^*\mathbb{R}$, thus identifying $\mathbb{R}$ as a subset of $^*\mathbb{R}$, (ii) tells us that $\mathbb{R}$ is a proper subset of $^*\mathbb{R}$. Elements of $\mathbb{R}$ are called *standard*, and elements of $^*\mathbb{R} - \mathbb{R}$ *nonstandard* real numbers. It can then be shown that among the nonstandard real numbers there exist ones that are infinitely large in that they exceed all the standard real numbers: the inverses of these (and their negatives) then constitute the infinitesimals of nonstandard analysis. If we consider the set $\mathbb{N}$ of (standard) natural numbers as a subset of $\mathbb{R}$, and hence also of $^*\mathbb{R}$, then it is easily seen that the set $\mathbb{I}$ of infinitesimals in the sense of nonstandard analysis is exactly

$$\{x \in^* \mathbb{R} : \forall n \in \mathbb{N}. -1/(n+1) < x < 1/(n+1)\}.$$

The members of $\mathbb{I}$ are thus cognate to the invertible infinitesimals present (as discussed above) in certain models of smooth infinitesimal analysis.

Much of the usefulness of nonstandard analysis stems from the fact that assertions involving infinitesimals (or more generally nonstandard numbers) are succinct translations to $^*\mathbf{U}$ of statements of standard classical analysis, involving limits or the '(ε, δ)' method. For example, let us say that x and y are *infinitesimally close* if $x - y$ is a member of $\mathbb{I}$. Then we find that the truth in $^*\mathbf{U}$ of the sentence

$$f(a+\eta) \text{ is infinitesimally close to } \ell \text{ for all infinitesimal } \eta$$

is equivalent to the truth in $\mathbf{U}$ of the sentence

$$\ell \text{ is the limit of } f(x) \text{ as } x \text{ tends to } \mathbf{a}.$$

And the truth in $^*\mathbf{U}$ of the sentence

$$f(a+\eta) \text{ is infinitesimally close to } f(\mathbf{a}) \text{ for all infinitesimal } \eta$$

is equivalent to the truth in $\mathbf{U}$ of the sentence

$$f \text{ is continuous at } \mathbf{a}.$$

Examples such as these show that nonstandard analysis shares with smooth infinitesimal analysis a concept of infinitesimal in which the idea of continuity is represented by the idea of 'preservation of infinitesimal closeness'.

Nevertheless, there are many differences between the two approaches. We conclude by stating some of them.

1. In models of smooth infinitesimal analysis, only smooth maps between objects are present. In models of nonstandard analysis, all set-theoretically definable maps (including discontinuous ones) appear.
2. The logic of smooth infinitesimal analysis is intuitionistic, making possible the nondegeneracy of the microneighbourhoods Δ and M_i, $i = 1, 2, 3$. The logic of nonstandard analysis is classical, causing all these microneighbourhoods to collapse to zero.
3. In smooth infinitesimal analysis, the Principle of Microaffineness entails that all curves are 'locally straight'. Nothing resembling this is possible in nonstandard analysis.
4. The property of nilpotency of the microquantities of smooth infinitesimal analysis enables the differential calculus to be reduced to simple algebra. In nonstandard analysis the use of infinitesimals is a disguised form of the classical limit method.
5. In any model of nonstandard analysis $^*\mathbb{R}$ has exactly the same set-theoretically expressible properties as $\mathbb{R}$ does: in the sense of that model, therefore, $^*\mathbb{R}$ is in particular an Archimedean ordered field. This means that the 'infinitesimals' and 'infinite numbers' of nonstandard analysis are so not in the sense of the model in which they 'live', but only relative to the 'standard' model with which the construction began. That is, speaking figuratively, a 'denizen' of a model of nonstandard analysis would be unable to detect the presence of infinitesimals or infinite elements in $^*\mathbb{R}$. This contrasts with smooth infinitesimal analysis in two ways. First, in models of smooth infinitesimal analysis containing invertible infinitesimals, the smooth line is non-Archimedean[48] in the sense of that model. In other words, the presence of infinite elements and (invertible) infinitesimals would be perfectly detectable by a 'denizen' of that model. And secondly, the characteristic property of nilpotency possessed by the microquantities of any model of smooth infinitesimal analysis (even those in which invertible infinitesimals are not present) is an intrinsic property, perfectly identifiable within that model.

The differences between nonstandard analysis and smooth infinitesimal analysis may be said to arise because the former is essentially a theory of infinitesimal numbers designed to provide a succinct formulation of the limit concept, while the latter is, by contrast, a theory of infinitesimal geometric objects, designed to provide an intrinsic formulation of the concept of differentiability.

[48] It is, however, smoothly Archimedean in the sense of exercise 8.2.

Appendix. Models for smooth infinitesimal analysis

In this appendix we sketch the construction of models for smooth infinitesimal analysis. We assume here an acquaintance with the basic concepts of category theory (see Mac Lane and Moerdijk, 1992 or McLarty, 1992).

The central concept in the construction of such models is that of a *topos*. To arrive at the concept of a *topos*, we start with the familiar category **Set** of sets whose objects are all sets and whose maps are all functions between them. We observe that **Set** has the following properties.

(i) It has a terminal object 1 such that, for any object X, there is a unique map $X \to 1$ (for 1 we may take any one-element set, in particular $\{0\}$). Maps $1 \to X$ correspond to elements of X.
(ii) Any pair of objects A,B has a *product* $A \times B$.
(iii) Corresponding to any pair of objects A, B there is an *exponential* object B^A whose elements correspond to arbitrary maps $A \to B$.
(iv) It has a *truth value* object Ω, containing a distinguished element *true*, with the property that for each object X there is a natural correspondence between subobjects of X and maps $X \to \Omega$. (In the case of **Set**, Ω may be taken to be any two-element **Set**, in particular the set $\{0, 1\}$, and *true* its element 1.) For any object X, the exponential object Ω^X then corresponds to the power set (object of all subobjects) of X.

All four of these conditions can be formulated in purely category-theoretic (i.e. maps only) language: a category satisfying them is called an (elementary) *topos*.

Associated with each topos **E** is a formal language $\mathscr{L}(E)$ called its *internal language* (see Bell, 1988b). This is a language which resembles the usual language of set theory in that among its primitive signs it has equality (=), membership ($\in$) and the set formation operator ($\{:\}$). It is, however, a many-sorted language, each sort corresponding to an object of **E**. Thus for each object

A of **E** there is a list of variables of sort A in $\mathscr{L}(\mathbf{E})$. Each term t of $\mathscr{L}(\mathbf{E})$ is then assigned an object B of **E** as a sort in such a way that, if t has free variables $x_1, \ldots, x_k$, of sorts $A_1, \ldots, A_k$, then t corresponds to a map $[\![t]\!] : A_1 \times \cdots \times A_k \to B$ in **E** called its *interpretation* in **E**. A *formula*, or *proposition*, is a term of sort Ω. A formula ϕ is said to be *true* in **E** if its interpretation $[\![\phi]\!]: A_1 \times \cdots \times A_k \to \Omega$ has constant value *true*. It can be shown that all the axioms and rules of inference of free intuitionistic logic (see Chapter 8) formulated in $\mathscr{L}(\mathbf{E})$ are true in this sense. Thus any topos is a model of intuitionistic logic.

One of the most important kinds of topos is obtained as follows. Start with any (small) category **C**. A *presheaf* on **C** is a functor[49] from $\mathbf{C}^{\mathrm{op}}$ – the opposite category of **C** in which all maps are 'reversed' – to **Set**. The presheaf category $\mathbf{C}^{\sim}$ has as objects all presheaves on **C** and as maps all natural transformations[50] between them. $\mathbf{C}^{\sim}$ is a topos; it is helpful to think of it as the topos of 'sets varying over **C**'. There is a natural embedding[51] – the *Yoneda embedding* – Y: $\mathbf{C} \to \mathbf{C}^{\sim}$, whose action on objects is defined as follows. For any object C of **C**, YC is the presheaf on **C** which assigns, to each object X of **C**, the set Hom(X, C) of maps in **C** from X to C. YC is the natural representative of C in $\mathbf{C}^{\sim}$, and the two are usually identified.

The concept of presheaf topos leads to the more general concept of *Grothendieck topos*, whose definition rests on the idea of a *covering system* (or Grothendieck topology) in a category. First, we define a *sieve* on an object C in a category **C** to be a collection S of maps in **C** satisfying $f \in S \Rightarrow f \circ g \in S$ for any map g composable with f. Equivalently, a sieve S may be regarded as a subobject or subfunctor of YC. Note that, if S is a sieve on C and h: $D \to C$ is any map with codomain C, then the set

$$h^*(S) = \{g\colon \mathrm{cod}\,(g) = D \text{ and } h \circ g \in S\}$$

is a sieve on D. Now we define a covering system on **C** to be a function J which assigns to each object C a collection $J(C)$ of sieves on C in such a way that the following conditions (i)–(iii) are satisfied. For any map f: $D \to C$, let us say that $S(J-)$ *covers* f if $f^*(S) \in J(D)$.

(i) If S is a sieve on C and $f \in S$, then S covers f.

[49] We recall that a *functor* between categories **C**, **D** is a function F that assigns to each object A of **C** an object FA of **D**, and to each map f: $A \to B$ of **C** a map Ff: $FA \to FB$ of **D** in such a way that $F1_A = 1_{FA}$ and, for composable maps f, g of **C**, $F(g \circ f) = Fg \circ Ff$.

[50] We recall that a *natural transformation* between two presheaves F and G on **C** is a function η assigning to each object A of **C** a map η_A: $FA \to GA$ of **C** in such a way that, for each map f: $A \to B$ of **C**, we have $G(f) \circ \eta_B = \eta_A \circ F(f)$.

[51] A functor F: $\mathbf{C} \to \mathbf{D}$ is called an *embedding* if, for any objects A,B of **C**, any map $FA \to FB$ in **D** is the form Ff for unique f: $A \to B$ in **C**.

(ii) *Stability*: if S covers $f: D \to C$, it also covers any composite $f \circ g$.
(iii) *Transitivity*: if S covers $f: D \to C$ and R is a sieve on C which covers all maps in S, then R also covers f.

We say that S *covers* C if it covers the identity map 1_c on C.

A *site* is a pair $(\mathbf{C}, J)$ consisting of a category $\mathbf{C}$ and a covering system J on it. Recall that any sieve S on an object C of $\mathbf{C}$ may be regarded as a subobject of the object YC of $\mathbf{C}^{\wedge}$. In particular, S may be considered an object of $\mathbf{C}^{\wedge}$. Now suppose we are given an object F of $\mathbf{C}^{\wedge}$ – a presheaf on $\mathbf{C}$ – and a map $f: S \to F$ in $\mathbf{C}^{\wedge}$ – a natural transformation from S to F. A map $g: YC \to F$ in $\mathbf{C}^{\wedge}$ is said to be an *extension* of f to YC if its restriction to the subobject S of YC coincides with f. We now say that the presheaf F is a (J-)*sheaf* if, for any object C, and any J-covering sieve S on C, any map $f: S \to F$ in $\mathbf{C}^{\wedge}$ has a unique extension to YC. Thus, figuratively speaking, a J-sheaf is an object F of $\mathbf{C}^{\wedge}$ which 'believes' that (the canonical representative YC of) any object C of $\mathbf{C}$ is 'really covered' by any of its J-covering sieves, in the sense that, in $\mathbf{C}^{\wedge}$, any map from such a J-covering sieve to F fully determines a map from YC to F.

It can then be shown that, for any site $(\mathbf{C}, J)$, the full subcategory $\mathbf{Shv}_J(\mathbf{C})$ of $\mathbf{C}^{\wedge}$ whose objects are all J-sheaves is a topos. Moreover, there is a natural functor $L: \mathbf{C} \to \mathbf{Shv}_J(\mathbf{C})$ called the *associated sheaf functor* which sends each presheaf F to the sheaf LF which 'best approximates' it. A topos of the form $\mathbf{Shv}_J(\mathbf{C})$ is called a *Grothendieck topos*. Models of smooth infinitesimal analysis are particular kinds of Grothendieck topos which will be described presently.

We say that a topos $\mathbf{E}$ is a *model* of **BSIA** if $\mathbf{E}$ contains an object R, together with maps $+: R \times R \to R$, $\cdot: R \times R \to R$ and elements $0, 1: 1 \to R$, for which the axioms of **BSIA** – expressed in the internal language $\mathscr{L}(\mathbf{E})$ – are true in $\mathbf{E}$.

To construct models of **BSIA**, we start with the category **Man**, whose objects are smooth manifolds and whose maps are the infinitely differentiable – smooth – maps between them. Now **Man** does not contain a microobject like the microneighbourhood Δ. Nevertheless we can identify Δ indirectly through its *coordinate ring* – that is, the ring $\mathbb{R}^{\Delta}$ of smooth maps on Δ to the real line $\mathbb{R}$. In fact, we know from exercise 1.8 that the coordinate ring $\mathbb{R}^{\Delta}$ should be isomorphic to the ring $\mathbb{R}^{*}$ which has underlying set $\mathbb{R} \times \mathbb{R}$ and addition $\oplus$ and multiplication $\otimes$ defined by $(a, b) \otimes (c, d) = (a + c, b + d)$ and $(a, b) \otimes (c, d) = (ac, ad + bc)$. This suggests that, in order to 'enlarge' **Man** to a category containing microobjects such as Δ – the first stage in the process of constructing models of **BSIA** – we first 'replace' each manifold M by its coordinate ring CM – the ring of smooth functions on M to $\mathbb{R}$ – and then 'adjoin' to the result every ring which, like $\mathbb{R}^{*}$, we want to be the coordinate ring of a microobject.

More precisely, we proceed as follows. Each smooth map $f: M \to N$ of manifolds yields a ring homomorphism $Cf: CN \to CM$ that sends each g in CN to the composite $g \circ f$: accordingly C is a (contravariant) functor from **Man** to the category **Ring** of (commutative) rings. We select a certain subcategory **A** of **Ring**, whose objects include all coordinate rings of manifolds, together with all rings which ought to be coordinate rings of microobjects such as Δ, but whose maps only include those ring homomorphisms which correspond to smooth maps. The contravariant functor C is then an embedding of **Man** into the opposite category $\mathbf{A}^{\mathbf{op}}$ of **A**. Thus $\mathbf{A}^{\mathbf{op}}$ is the desired enlargement of **Man** to a category containing microobjects, and only smooth maps. However, $\mathbf{A}^{\mathbf{op}}$ is not a topos, so we need to enlarge it to one. The natural first candidate presenting itself here is the presheaf topos $\mathbf{Set}^{\mathbf{A}}$, with the Yoneda embedding $Y: \mathbf{A}^{\mathbf{op}} \to \mathbf{Set}^{\mathbf{A}}$. The composite $i = Y \circ C$ then embeds **Man** in $\mathbf{Set}^{\mathbf{A}}$. In the latter the role of the smooth line R is played by the object $i(\mathbb{R})$, and that of Δ by the object $Y(\mathbb{R}^*)$ – the images in $\mathbf{Set}^{\mathbf{A}}$ of $C\mathbb{R}$ and $\mathbb{R}^*$, respectively.

Now $\mathbf{Set}^{\mathbf{A}}$ is 'almost' a model of **BSIA**. This is because the truth of most of the axioms of **BSIA** in $\mathbf{Set}^{\mathbf{A}}$ can be shown to follow from certain facts about $\mathbb{R}$ considered as a (classical) object of **Man**, or its coordinate ring $C\mathbb{R}$ as an object of **A**. For instance, the assertion 'R is a commutative ring with identity' follows from the corresponding fact about $\mathbb{R}$. The correctness of the Principle of Microaffineness for $\mathbb{R}$ in $\mathbf{Set}^{\mathbf{A}}$ can be shown to follow from a result of classical analysis known as Hadamard's theorem, which asserts that, for any smooth map $F: \mathbb{R}^n \times \mathbb{R} \to \mathbb{R}$, there is a smooth map $G: \mathbb{R}^n \times \mathbb{R} \to \mathbb{R}$ such that

$$F(x, t) = F(x, 0) + tF_t(x, 0) + t^2G(x, t).$$

(That is, modulo t^2, any smooth map $F(x, t)$ is affine in t.) Unfortunately, however, certain key principles of **BSIA** are not true in $\mathbf{Set}^{\mathbf{A}}$. For example, the assertion that $\forall x \in R(x < 1 \vee x > 0)$, which may be paraphrased as 'the intervals $(\leftarrow, 1)$ and $(0, \to)$ cover R', is false in $\mathbf{Set}^{\mathbf{A}}$, although the corresponding principle for $\mathbb{R}$ is evidently true. This may be summed by saying that the embedding i fails to preserve open covers.

This deficiency is rectified by imposing a suitable covering system on $\mathbf{A}^{\mathbf{op}}$ and then considering the topos of sheaves with respect to this covering system. This has the effect of cutting down $\mathbf{Set}^{\mathbf{A}}$ to those presheaves on $\mathbf{A}^{\mathbf{op}}$ which 'believe' that open covers in **Man** are still open covers in $\mathbf{Set}^{\mathbf{A}}$. It can then be shown (see Moerdijk and Reyes, 1991) that the resulting topos **E** of sheaves is a model of all the principles laid down in **BSIA**. In **E**, LR is the smooth line and $L\Delta$ the principal microneighbourhood of 0, where $L: \mathbf{Set}^{\mathbf{A}} \to \mathbf{E}$ is the associated sheaf functor.

Suitable refinements of the choice of covering system in $\mathbf{A}^{\mathrm{op}}$ lead to toposes of sheaves which can be shown to satisfy the other principles of smooth infinitesimal analysis discussed in the text. For each such topos **E** we have a chain of functors

$$\mathbf{Man} \overset{C}{\to} \mathbf{A}^{\mathrm{op}} \overset{Y}{\to} \mathbf{Set}^{\mathbf{A}} \overset{L}{\to} \mathbf{E}$$

whose composite s: **Man** $\longrightarrow$ **E** can be shown to have the following properties.

$s(\mathbb{R})$ = the smooth line R.
$s(\mathbb{R} - \{0\})$ = the set of invertible elements of R.
$s(f') = s(f)'$ for any smooth map $f: \mathbb{R} \to \mathbb{R}$.
$s(TM) = (sM)^{\Delta}$ for any manifold M, where TM is its tangent bundle in **Man**.

Such models of **BSIA** are said to be *well adapted.*

Let us call the image $s(M)$ of a classical manifold M in a well-adapted model **E** of **BSIA** its *representative* in **E**. At first it might be thought that classical manifolds and their representatives are radically different. For example, classically, for any point x on the real line $\mathbb{R}$, either $x = 0$ or $x \neq 0$, but as we know, this is not the case for its representative the smooth line R. However, this difference is less deep than it seems. In fact, if we analyse the meaning of any statement of the internal language of **E** containing a variable over R, we find that it is true in **E** if and only if the corresponding statement in **Man** is, in addition to being true for all points of $\mathbb{R}$, is also locally true for all smooth maps to $\mathbb{R}$. A smooth map $f: \to \mathbb{R}$ may have $f(a) = 0$ for some point a and yet not be constantly zero on any neighbourhood of a. This means that neither $f = 0$ nor $f \neq 0$ is locally true at a, so that '$f = 0 \vee f \neq 0$' fails to be locally true. Similarly, the trichotomy law $\forall x \forall y\ (x < y \vee x = y \vee y < x)$, although true in $\mathbb{R}$, fails for R, since for smooth maps $f, g: \mathbb{R} \to \mathbb{R}$ there may be points a on no single neighbourhood of which do we have $f < g$ or $f = g$ or $g < f$. On the other hand, since f is continuous, each point a either has a neighbourhood on which $0 < f$ or one on which $f < 1$, so that the statement '$0 < x \vee x < 1$' holds for every x in R.

Most well-adapted models **E** of **BSIA** have the further property that elements of the smooth line R, that is, maps $1 \to R$ in **E**, correspond to points of $\mathbb{R}$ in **Man**. (This means that in passing from $\mathbb{R}$ to R no new 'genuine' elements are added, but only 'virtual' ones.) As we remarked in Chapter 8 (see McLarty, 1988), it can also be shown that these models satisfy the closed law of excluded middle in the sense that $\alpha \vee \neg\ \alpha$ is true whenever α is a closed sentence, that is, one devoid of free variables. In particular, for any elements a,b of R the statement

$$a = b \vee a \neq b$$

is true. But of course we cannot go on to infer from this that

$$\forall x \in R \forall y \in R[x = y \vee x \neq y],$$

hence we know from Theorem 1.1(iii) that this assertion is false in any smooth world. Thus R is, unlike an ordinary set, more than the mere 'sum' of its elements. This is in fact a typical feature of objects in a topos: for details see Bell (1988b).

Note on sources and further reading

The programme of developing the concept of smoothness in category-theoretic terms and of reviving the use of nilpotent infinitesimals in the calculus and differential geometry was first formulated by F. W. Lawvere in lectures delivered in Chicago in 1967 (Lawvere, 1979; 1980), it is to be noted that a central motivation for this programme (and for the subsequent development of topos theory) was to furnish an adequate axiomatic framework for continuum mechanics. These 1967 lectures contain the first constructions of toposes realizing the principle of Microaffineness in the sense of containing an 'infinitesimal' object Δ for which R^{Δ} is isomorphic to $R \times R$. Here are also to be found the identification of S^{Δ} as the tangent bundle of a space S and the equivalence of vector fields, microflows and microtransformations (as presented in Chapter 7). The Principle of Microaffineness in the explicit form given in Chapter 1 (i.e. with a specified isomorphism between R^{Δ} and $R \times R$: see exercise 1.6) was introduced by Kock (1977) and is often referred to as the Kock–Lawvere axiom. The investigation of toposes realizing the Principle of Microaffineness and into which the category of manifolds can be 'nicely' embedded (the so-called 'well-adapted' models) was first carried out in Dubuc (1979). The first systematic account of synthetic differential geometry was given by Kock (1981): I have drawn on the first part of this work extensively. Another useful work is Lavendhomme (1987, 1996), which provides an elegant axiomatic development of the subject. There is also a brief introductory account in McLarty (1992). The major work on the actual construction of topos models for synthetic differential geometry is the book by Moerdijk and Reyes (1991).

Elementary approaches to the smooth world include McLarty (1988), Bell (1988a) – a principal source for this book – and Bell (1995).

The book by Lawvere and Schanuel (1997) contains an introduction to category theory aimed at beginners. More advanced treatments include Mac Lane (1971) and McLarty (1992). The literature now includes several books on topos theory. The most elementary of these is Goldblatt (1979), and the most advanced Johnstone (1979). Somewhere in between are Barr and Wells (1985), Bell (1988b), Freyd and Scedrov (1990), Lambek and Scott (1986), Mac Lane and Moerdijk (1992), and McLarty (1992). Bell (1986) is an attempt to formulate some of the philosophical ideas suggested in the emergence of topos theory.

For nonstandard analysis see Robinson (1966) and Bell and Machover 1977).

For intuitionistic or constructive logic see Bell and Machover (1977), Dummett (1977) and Kleene (1952).

For an account of the development of the concepts of the continuum and the infinitesimal see Bell (2005a), (2005b).

Finally, for the history of the calculus I have found the books of Baron (1969) and Boyer (1959) most useful. Some of the applications in Chapter 4 have been adapted from Gibson (1944).

References

Aristotle (1980). *Physics*, Vol. II. Cambridge, MA: Harvard University Press.
Banach, S. (1951). *Mechanics* (trans. E. J. Scott). Warszawa: PWN.
Baron, M. E. (1969). *The Origins of the Infinitesimal Calculus*. Oxford: Pergamon Press.
Barr, M. and Wells, C. (1985). *Toposes, Triples and Theories*. Berlin: Springer-Verlag.
Bell, J. L. (1986). From absolute to local mathematics. *Synthese*, **69**, 409–26.
Bell, J. L. (1988a). Infinitesimals. *Synthese*, **75**, 285–315.
Bell, J. L. (1988b). *Toposes and Local Set Theories*. Oxford: Clarendon Press.
Bell, J. L. (1995). Infinitesimals and the continuum. *Mathematical Intelligencer*, **17**(2), 55–7.
Bell, J. L. and Machover, M. (1977). *A Course in Mathematical Logic*. Amsterdam: North-Holland.
Bell, J. L. (2005a). *The Continuous and the Infinitesimai in Mathematics and Philosophy*. Milano: Polimetrica.
Bell, J. L. (2005b). *Continuity and Infinitesimals*. Stanford Encyclopedia of Philosophy.
Boyer, C. B. (1959). *The History of the Calculus and its Conceptual Development*. New York: Dover.
Brouwer, L. E. J. (1964). Intuitionism and formalism. In *Philosophy of Mathematics, Selected Readings*, eds P. Benacerraf and H. Putnam. Oxford: Blackwell.
Cascuberta, C. and Castellet, M., eds (1992). *Mathematical Research Today and Tomorrow: Viewpoints of Six Fields Medallists*. Berlin: Springer-Verlag.
Courant, R. (1942). *Differential and Integral Calculus*. London: Blackie.
Dubuc, E. (1979). Sur les modeles de la geometrie differentielle synthetique. *Cahiers de Topologie et Geometrie Differentielle*, **XX-3**, 231–79.
Dummett, M. (1977). *Elements of Intuitionism*. Oxford: Clarendon Press.
Freyd, P. J. and Scedrov, A. (1990). *Categories, Allegories*. Amsterdam: North-Holland.
Gibson, G. (1944). *An Introduction to the Calculus*. London: Macmillan.
Goldblatt, R. I. (1979). *Topoi: The Categorial Analysis of Logic*. Amsterdam: North-Holland.
Heyting, A. (1971). *Intuitionism: An Introduction*. Amsterdam: North-Holland.
Hohn, F. E. (1972). *Introduction to Linear Algebra*. New York: Macmillan.
Johnstone, P. T. (1979). *Topos Theory*. London: Academic Press.
Kant, I. (1964). *Critique of Pure Reason*. New York: Macmillan.
Kleene, S. C. (1952). *Introduction to Metamathematics*. Amsterdam: North-Holland and New York: Van Nostrand.
Kock, A. (1977). A simple axiomatics for differentiation. *Mathematica Scandinavica*, **40**, 183–93.

Kock, A. (1981). *Synthetic Differential Geometry.* Cambridge: Cambridge University Press. (Second edition, 2006)

Lambek, J. and Scott, P. J. (1986). *Introduction to Higher-Order Categorical Logic.* Cambridge: Cambridge University Press.

Lavendhomme, R. (1987). *Lecons de Geometrie Synthetique Differentielle Naive.* Louvain-La-Neuve: Institut de Mathematique.

Lavendhomme, R. (1996). *Basic Concepts of Synthetic Differential Geometry.* Dordrecht: Kluwer.

Lawvere, F. W. (1979). Categorical dynamics. In *Topos Theoretic Methods in Geometry*, Aarhus Math. Inst. Var. Publ. series 30.

Lawvere, F. W. (1980). Toward the description in a smooth topos of the dynamically possible motions and deformations of a continuous body. *Cahiers de Topologie et Geometrie Difféentielle*, **21**, 377–92.

Lawvere, F. W. and Schanuel, S. (1997). *Conceptual Mathematics: A First Introduction to Categories.* Cambridge University Press.

Mac Lane, S. (1971). *Categories for the Working Mathematician.* New York: Springer-Verlag.

Mac Lane, S. and Moerdijk, I. (1992). *Sheaves in Geometry and Logic: A First Introduction to Topos Theory.* New York: Springer-Verlag.

McLarty, C. (1988). Defining sets as sets of points of spaces. *Journal of Philosophical Logic*, **17**, 75–90.

McLarty, C. (1992). *Elementary Categories, Elementary Toposes.* Oxford: Clarendon Press.

Misner, C., Thorne, K., and Wheeler, J. (1972). *Gravitation.* Freeman.

Moerdijk, I. and Reyes, G. E. (1991). *Models for Smooth Infinitesimal Analysis.* New York: Springer-Verlag.

Peirce, C. S. (1976). *The New Elements of Mathematics*, Vol. III, ed. C. Eisele. Atlantic Highlands, NJ: Humanities Press.

Rescher, N. (1967). *The Philosophy of Leibniz.* Englewood Cliffs, NJ: Prentice-Hall.

Robinson, A. (1966). *Non-Standard Analysis.* Amsterdam: North-Holland.

Russell, B. (1937). *The Principles of Mathematics*, 2nd edn. London: George Allen and Unwin Ltd.

Spivak, M. (1979). *Differential Geometry*, 2nd edn. Berkeley: Publish or Perish.

Van Dalen, D. (1995). Hermann Weyl's intuitionistic mathematics. *Bulletin of Symbolic Logic*, **1**(2), 145–69.

Wagon, S. (1985). *The Banach–Tarski Paradox.* Cambridge: Cambridge University Press.

Weyl, H. (1921). Uber die neue Grundlagenkrise der Mathematik. *Mathematische Zeitschrift*, **10**, 39–79.

Weyl, H. (1922). *Space–Time–Matter.* New York: Dover.

Weyl, H. (1940). The ghost of modality. In *Philosophical Essays in Memory of Edmund Husserl.* Cambridge, MA: Harvard University Press.

Weyl, H. (1987). *The Continuum: A Critical Examination of the Foundation of Analysis* (transl. S. Pollard and T. Bole). Philadelphia: Thomas Jefferson University Press.

Index